Beekeeping

How to Avoid Common Mistakes and Pitfalls

(A Step-by-step Guide to Beekeeping for Beginners)

Jonathan Hudson

Published By **Simon Dough**

Jonathan Hudson

All Rights Reserved

Beekeeping: How to Avoid Common Mistakes and Pitfalls (A Step-by-step Guide to Beekeeping for Beginners)

ISBN   978-1-998927-46-3

No part of this guidebook shall be reproduced in any form without permission in writing from the publisher except in the case of brief quotations embodied in critical articles or reviews.

Legal & Disclaimer

Table Of Contents

Chapter 1: Honey Bees: All you need to know 1

Chapter 2: How to Get started with beekeeping 33

Chapter 3: Maintaining your Honeybees 90

Chapter 4: Bee Harvests........................ 144

Chapter 5: Beekeeping Supplies........... 167

Chapter 6: Preparation and storage of the Bee Boxes 171

Chapter 7: Safety with bees 174

Chapter 8: The Beginning of the Bees .. 176

Chapter 9: How to Maintain the Hives . 179

Chapter 10: The Harvesting of Honey .. 182

Chapter 1: Honey Bees: All you need to know

One of the best ways to be a good beekeeper is to understand the animals that you are working with. Bees have become more important to humans over the years. Because of their vulnerability, understanding how they work in our lives will be a huge help for beekeepers.

This part will cover the physiological changes occurring in honeybees based on their position, breed, and sex. Learn about bees' importance for human endurance, as well how they can help the state. It's amazing to see how 40,000 bees work together for common objectives, all in the interest of the overall wellbeing of the province.

Honeybees: Why we need them

Honeybees can do much more than make honey, despite the fact that they are called honeybees. They are responsible for 33% global nourishment. This is comparable to the flexibility of plants and trees. Aside from that, they could be the best examples of superorganisms to help Earth's animals, such as humans.

Every year, honeybees fertilize more US crops than $15 billion. For the fertilization of more than 90 unique US crops, including almonds, horses feed seed, vegetable and apple seeds, beekeepers rent over 2.5 million settlements annually. These yields cover approximately 3.5million acres and represent 33% to our daily diet. Honeybees also ensure that those nutrients stay in plentiful, gracefully. Because bees are usually dynamic, it doesn't matter when the season is, people can rest assured that we will receive nourishment year round.

Bees fertilize flower blossoms better than any creepy-crawly, and better than any other plant. Surprisingly bee fertilization is dependent on nutrition to thrive. Working drones move from blossom to blossom while drinking nectar. As she does so, she encounters dusty anthers, which can run to her hairs. As she glides along, she begins brushing dirt with her legs and moving it into the dust bins (CORBICULA), on her rear leg. During this process, she accidentally dropped some of her dust onto other blossoms, which caused fertilization.

Bees have superpowers

The province works together to manage the hive temperature and move air around and throughout the hive, to cool it, heat and acquire more oxygen. They can sometimes travel up to 6 miles per day on one excursion. They gather water and manage dampness. The sharing of hormonal emission and sovereign substances among bees gives them a whole new level of control over their minds.

The Honeybee Superorganism

Because honeybees are social animals, they can survive alone for long periods of time. A province of bees is also known as a superorganism. Honeybees have the ability as an aggregate to live together and cooperate within a state. While each bee goes through different stages of life and has specific tasks to complete, overall, province prosperity determines the bee's behavior.

About 80% percent of plants use bio fertilization. This means that they require the assistance of other living organisms (such as creepy crawlies and creatures) in order to move the soil. However, different plants like grasses, conifers, deciduous and deciduous tree use abiotic fertilization.

Regular Honeybees

Some bees may not be the same, and some bees don't produce honey. Honeybees are the name they use to describe themselves. This is because it's all about sweet nectar, fertilization, and how they reproduce. There are many types of honeybees. Each has its own attributes and practices.

The Western Honeybee

While the honeybee variety includes some animal types, this book focuses on one of the most well-known honeybees on the planet: The Western honeybee. You would first have found this species in Africa, Europe or Asia. In any event, many subspecies or geographic race began to emerge when people started to import bees to different geographic areas.

Because of its flexibility, honey-creating capabilities, and friendly boss interaction, the Western honeybee remains the most recognized bee. As beekeeping became more common, beekeepers began searching for bees based on their most desirable qualities. This included shading, size and reasonableness to a specific area. They also looked at personality, swarm affinity, propolis (tarbees gather) and protective practices. This led to specific characteristics being tested for in bee raising. Many half breeds like Buckfast, Starline or Midnight were created by interbreeding to improve their attractive characteristics. Even though only a fraction of the half and halves, such as Starline, Midnight and Cordovan, is currently accessible, others are still being tested and created in serious reproducing programs throughout the United States.

US beekeepers have yet to recognize Carniolan or Italian bees as the most commonly purchased honeybees. However with new research focusing on bee illnesses and parasites, science may soon be able to raise more grounded bees with greater resistance and opposition. This should result in honeybees being able to continue their flourishing and providing the world with their fertilization enchantment.

The Honeybee Hive

While the honeybee bee hive has many complex signs and structures, it is still a system that is quite mind-blowing. The purpose of existence is to create conditions where the province can flourish and survive. The honeybees need to find and maintain an adequate area to build their home of wax. They also need it for the purpose of feeding the brood. They look for a home area that is within a cavity. This could be in a tree in the ground or a man-made structure. The two most important requirements for survival in a settlement are:

1) You need to find a suitable place for bees and beekeepers to gather.

2) Instill hereditary characteristics through automaton creation, conceptive and throwing swarms.

Wild bees have a tendency to seek out home. These places need to be cozy so that they can safeguard and secure their colonies. However, they also need space to build and store honey. They often select locations that were previously used by different honeybees. It's possible for bees to make their homes anywhere they like.

Once a province has chosen an area, the bees immediately go to work fixing any breaks or covering the internal partitions with propolis. This helps weatherproof the colony and also provides antimicrobial as well antifungal properties that help keep the hive stable. They also work in the colony to maintain the appropriate temperatures for brood rearing. This may include cooling or warming the hive depending upon temperature.

The honeybees will begin to put together their brushes once they have selected their new home. This involves joining four to eight brush ends to the roof or sides of the house. Bee space refers to the fact that all brushes are made in a similar way and each one has a distance of around 1 cm. This allows bees to move freely around the home and not worry about finding one another.

Making honeycomb is an extremely vitality-consuming job, so honeybees try to be successful in their endeavors. Because they are often starting without preparation, their first year in another region is usually the most difficult. They'll need lots honey stores to construct their house. However, they'll also need enough honey to sustain them for their first winter.

The movement in the beehive does not stop. Bees are always busy. They build, clean, evaluate, secure, take care of the sovereign, her brood, search for more food sources, and they also look after their own health. They are no surprise to be the most common source of the expression "occupied, as a Bee".

Phases of Honeybee Development

As with other creepy crawlies honeybees also experience improvements from childhood to adulthood. Their lifespans are dependent on their capacities, but it's the sovereign and what eggs she lays who decide every bees job -- and its destiny.

EGG

Working drones can survey the needs of a state and then assemble wax cells of various sizes to address these issues. The sovereign then can quickly identify the format of each cell and determine which kind of egg to lay. A prepared egg results in a female honey bee, while unfertilized eggs result in a male honey bee. This egg stage can last for three days.

Female bees can be diploid which means they inherit two sets of chromosomes from their parents. Male bees have one lot of genes from their mom, but are haploid. This genetic cosmetics won't likely mean much to the sovereign. However, this becomes an important aspect for the eggs she lay.

Hatchling

After three days, an egg hatches into the first instar. The hatchling will be laid in a C-formed position at the base. The hatchling is dependent on ample amounts of nutrition and begins to develop at multiple times its normal size each day. The new hatchlings get a rich, nutritious diet that includes illustrious jam. If the hatchling is going to be a monarch, she will continue her strict diet of illustrious jelly. However, automatons and labourers can change to feed their brood with a mixture of dust and honey. Being a good parent is hard work. Every hatchling will need to be visited at least once in the first 8-10 days.

Honeybees' Life Cycle

A province that runs a mill has anywhere from 20,000 to 60,000 honeybees. However, a greater proportion of the bees is female working drones. Depending upon the season, and more importantly, the state's prosperity, a provincial may have approximately 300 male bees called "rambles". Be that though, it is possible for a settlement to have only one sovereign.

This is because she is the mother to the many bees living in the province. Every one of these bees have a new job and an outstanding burden in the hive. And everyone has an alternative life cycle.

PUPA

After being released into the world eight days ago, labourer hatchlings are closed in their cells. On the ninth day, hatchlings are covered with silk using a glandular emitter in their brains. To finish their prepupal stage on the tenth day the hatchlings are instructed to lie back and face the cell's highest point. The pupae become whiter after the fifth shed is completed on the 11th day. The bees increase their shading week by week and will eventually experience their sixth and last home. This is when bees move from pupae stage to imago and begin to resent their own behavior.

Each of the three positions/phases bees--sovereigns, automatons or labourers--starts with the same 3-day embryo stage. They differ, with the sovereign being the shortest of the larval and pupal stages while the automaton the longest. A critical indicator device for beekeepers is the honeybee's life stages table. This is used to assess hives looking for great sovereigns with excellent laying designs. It is important to consider all factors involved in the raising of eggs, such as diet and preparation. This will give you an immediate indication of what kind of adult rises from the cell, how long it lives, and what bigger role it plays inside the hive.

The Queen Bee

Stable and happy sovereigns are essential to the endurance of bees. The sovereign manages the settlement, employs labourers and automatons, and she still manages to lay eggs every day. However, she is more than a mothering sovereign.

Turning into a Queen

Any honeybee-treated egg can be converted into a workhorse or a sovereign. This is called totipotent. Worker drones begin to search for young hatchlings and provide them with an excellent eating program rich in honey, dust, and other bee catalysts. This more demanding eating regime allows her to have a bigger stomach and more ovarioles, which are cylinders to her ovaries. She is currently on her way towards becoming a sovereign.

To prevent a sovereign from leaving the hive, some attendants may remove her sovereign's wing. However, if her wing gets cut by the province and she attempts a escape, she may be forced to land on the ground. If another little girl becomes the sovereign in the settlement, the new girl can execute the sovereign.

Sovereign Cups or Cells

Anywhere in the beehive, sovereign cups or cells are possible. They're not permitted to be placed in brood brushes. Sovereign cup is the beginning of an sovereign cell. But, they are not equipped with illustrious jelly, eggs, or hatchlings. Numerous states possess at least one sovereign cup. This is to protect them from the possibility of making a sovereign. A sovereign cup is a cup with a longer neck that has been cleaned and polished. It also has regal jam on its base. A sovereign can lay twice as many eggs each day as her body weight, and she will keep delivering eggs every single day for a remarkable remainder, which can last around 3-4 years. Every day during spring, the sovereign lays as many as 2,000 eggs, pressing cells rapidly with very small mouths.

The sovereign is not hard to spot. She waddles across the brush with apparent purpose, scanning for newly arranged or empty cells in which she can store eggs. She can have many hues on her midriff as well as groups, but she is most often a strong shading with no stripes. Depending on the place you purchased your sovereign, you can ask the raiser for a shading spot to be placed on her back that is related to her world year. Also, they may request to slit one wing to keep her safe from flying in a multitude.

In order for the sovereign to continue to be the main egg layer in the honeybee hive she should send pheromones to the working drones. QMP (QUEEN MANDIBULAR pheromone), which is produced in the sovereign's mind, is one of the major pheromones. QMP suppresses the potential for ovary enhancement in the labourers. It also encourages the formation of specialist entourages and postpones searching. In a matter of days or seven days, a virgin sovereign, newly incubated, leaves her home to go on mating trips in ramble assembly regions. Here she flies around while holding on to being trapped in mid-air and mated in rambles. She makes a variety of mating trips each time before returning to her home. There, she'll lay eggs for the rest of her life.

These flights see her mate with many automatons in order to get enough spermatozoa, which treats the ovum, in her Spermatheca (where sperm are put away). A few of the spermatozoa blends are released when the sovereign releases eggs to create future working drones. These working drones consist of a number of relatives each with the varied qualities and attributes of their respective dad rambles. The decent automaton range takes into account greater province endurance, as certain automatons may have higher-scrounging and other malady resistance.

The Drone Bee

Automaton honeybees, the main guys of a state, don't accumulate dust or nectar. And because they don't have bee stingers they can't help protect the hive. Ramble bees only have one job in their daily lives: to mating with virgin sovereigns to give birth to their hereditary traits.

When a settlement is moving well and becomes more prominent, with lots of honey, your bees could go into conceptive multitude mode. This is when the drones in action start building larger cells to cover the brush edges and fill the empty spaces below or under the edge. The larger cells are an indication that the sovereign intends to lay unfertilized ovules, which will ultimately become her children: the automatons. A settlement will typically have between two and three hundred automatons. Rambles are more prominent in a settlement, so they use more of the hive resources. Automatons have haploid status, meaning that they are as they were. This automaton will be taken care by working drones. For mating, they also have larger flight muscles. In winter, the drones will work to remove any rambles and other debris from the settlement.

How drones mate with queens

To mate with the virgin sovereign, an automaton puts his endophallus in a sovereign. The penis stays inside the flying sovereign. Now, the automaton is thrown to the ground, and it continues to eat the dust. The next automaton to retrieve the sovereign will expel penis and rehash procedure, allowing the sovereign start to fill her infected oviducts hereditarily diverse spermatozoa. Instead they mate in ramble assembly territories that are 30 to 60 feet wide. These identical zones are frequently used over a period of time. However, it's unclear how new automatons can find these territories and how they can be controlled by other automatons. Automatons may be located within 300 feet of anapiary once in a while, although they are more common to be far away. Automatons that form an assembly zone can come from upwards to 200 different provinces. Additionally, up to 25,000 automatons are allowed to make up one automaton-gathering territory. This expands decent hereditary variety while decreasing the chance of inbreeding.

In what degree do Drones Live

Automatons are able to wander from hive in a flurry of activity, and can spend their evenings making 30-minute to 1-hour trips across automaton assembly territories looking for virgin sovereigns. Although one automaton per 1,000 is successful in his mating mission, there are far more sovereigns than automatons. These unproductive automatons can usually be satisfied for half a century. Effective automatons, which can be as amusing and entertaining as that sounds, are usually able to live for a half year.

The Worker Bee

They perform the day-to-day tasks that keep the beehive in good shape, just like their name. They continuously assess the hive, look for opportunities to finish projects and move in the same direction as their hive sisters. You shouldn't dismiss them from their settlement positions. They are the very core of the state. Without them, automatons and sovereigns would cease to exist.

Turning into a Worker

It takes about 21 days for a functioning drone to be born from the egg. All drones that are working are females, and they were born from ready-to-use eggs. They get most of their hereditary cosmetics and half of them from their automaton mother. Working drones in the province are always made up of both full and stepsisters because the sovereign has a lot of automaton mates and mixes all her sperm with her spermatheca. This gives them a larger range of hereditary abilities. Some may be better at creating wax, others will be dust foragers, and some will be honey creators. All of them are contributing their solidarity to building settlement endurance.

The list of occupations for drone workers is extensive. They begin their work right after they are born. Their first task is cleaning out the world cell where they were introduced. For the first three weeks, house bees work inside the hive. These include feeding and caring for the young hatchlings, topping cells, maintaining their mother, cleaning out the world cell, maturing honey, protecting the queen, and much more. Once more bees become incubated and are capable of taking over house jobs, the labourers can begin to fly direct. These short round trips take you up, around, and back to your hive. They're learning about their environment, and are making a map of it so they can return to the hive when they need food.

Bees need to survive by scavenging. Forager bees seek out food for the hive and will spend their entire life as tracker gatherers. They'll search within a 2-3 mile radius of the hive for nectar, water, propolis, and dust. Forager bees are known to fly as far as 6 miles in search for nectar and dust when resources in an inquiry area are low.

The Beekeeper's Notebook

Healthy wing flight time for a bee is about 500 miles. Keep your bees close to food and water sources so they can live longer. The point at which temperatures drop and daylight hours diminish, bees are less likely to fly and will live for half as long.

It can be difficult to decide a drone's life expectancy. But it is dependent on average temperatures. They may live 2 to 3 days during periods when the nectar stream is abundant and they are speeding out. They are not ready to fly back into their hives and they will eventually die from the relentless flying.

When you observe a whirlwind bees fly around your hives you might think they are in trouble. But together they are a finely tuned machine. The bee rank framework of the sovereign and labourers signifies every bee's individual capabilities.

This is because bees are able to understand what they need to achieve in order to enable the hive function productively.

How Bees Communicate

Bees are a superorganism. They rely on one another to survive; being open is key for successful execution. Bees monitor the state of each hive to ensure that they are profitable and secure their future. Most correspondence takes place using pheromones. They travel around the settlement to distinguish each other with their intense sense of smell. Trophallaxis, which is the act of taking care practices, also makes them release pheromones.

Where is King Bee located?

There is no ruler honey in the hive. Initialy, the sovereign honey was called the ruler bee due to its larger size and greater significance in the province. After her death, scientists discovered she was a female and that she was the egg-layer of the settlement. Even though they don't do much outside the hive but lay eggs, the soleign is the most important job a honey bee can have. Her egg-laying ability determines the success or failure of all future bees.

These drones work in the hive and are called working drones. They agree on a few important decisions such as the availability of hive space and search nourishments, climate conditions and the wellbeing of the hive. Automatons aren't equipped with stingers and do not gather nutrition for the hive. They have one primary job: to mate and mat with virgin sovereigns from other hives. Automatons may live for up to a year and a bit, but they soon die when they mate. Similar to humans, bees communicate with one another in a variety of ways. Bees can share data with others through both physical and compound signals: vibration, contact (smell), taste, smell, development, vibration, touch and contact. All these signs are intended to help different bees find food, regardless of their risk.

Bee Senses

Honeybees perceive their surroundings using different tangible upgrades. In addition to being able to provide information to their colonies regarding food, dangers, or other situations, they also have an amazing capacity to pass on mind-boggling subtleties. They are able to use their inner GPS to determine their environment factors and to communicate the types and locations of their food sources. Bees are sensitive and sensitive to tastes and smells. They use this information to find nectar sources and dust sources when they rummage. They can see all hues except the shading, which to bees looks dark. Honeybees also have the ability to discern brighter ranges, which aids in their search for food. They can detect if a bloom has been recently visited by another honeybee and will therefore avoid it.

What Bees Eat

The kind of food honeybees can find and store has a huge impact on many aspects of province life. How bees eat can impact their health, development and longevity, as it can affect their invulnerability and ability to survive. It might appear that bees produce honey to be used by others. However, bees depend on honey and other nutrients they grow and make their own food. They won't be able take care their posterity or themselves if they can't access water, dust, or nectar.

HONEY

Bees turn nectar gathered from flowers into honey, which is their primary source of starch. They also care for developing hatchlings. The hatchlings chosen to become future sovereigns get royal jam. All bees enjoy honey as a source of vitality. It is also an important food source for the bees in colder months. People also love honey but try to leave a bounty for your bees' nourishment.

BEE BREAD

Together with honey, nectar and bee salivation bees transform dust into matured nourishment. This provides bees with the protein they require. As dust can't be stored for too long after it's been expelled from plants, bees mature it to increase its lifespan. This also allows the bees to eat decent food during cold months when they aren't able or unable to collect honey or dust. All hatchlings are treated with regal jelly. After three days, however regal jam is removed from all hatchlings.

Dust

The dust is used as a protein source in bee bread. In addition to being consumed by hatchlings, it's also consumed by medical attendants bees for their physical development. These bees' young glandular development is dependent on dust. It is used by the organs of their organs (hypopharyngeal as well as mandibular) to produce imperial jelly. A decent amount of dust can be made by labourers, which will also provide all the minerals and nutrients necessary to keep the settlement safe and sound.

NECTAR

Nectar provides the sugars and vitality bees need to fly or work. To take care of their young bees, they blend it with some dust. They store the honey at the edges as topped honey during winter. They need to make honey, and they must also use it to create a defensive bunch around their sovereign.

WATER

Water is also a crucial component of honeybee food. They can't go more than a few days without drinking water. They need water to absorb the supplements they eat, and it is necessary to remove squander. Water is also used for condensing thick honey and sugar. This makes it more appealing to be eaten. Water should be kept near the beeyard. Water can be placed in feeders at the hive for extreme weather conditions.

Chapter 2: How to Get started with beekeeping

Beekeeping has moved from being a hobby to an artistic outlet. In the last 200 years, thousands of mechanical innovations have been made that have allowed us to stay at the forefront of beekeeping.

The last couple of years have seen a renewed interest in specialist beekeepers, who may need to own a few settlements for some honey or simply to help the bees. This section covers the most important information in beekeeping history, as well the types of hives. In this section, you will also learn about the tools and equipment required to start beekeeping. You will also learn how to build your beard.

Evolution of Beekeeping

Although beekeeping might seem to be a fringe interest, the fact is that beekeeping has been a long-standing hobby. Honey has been made for thousands of years. It was not possible to find the same innovative methods for beekeeping as we do today. However, they were able to quickly discover how to keep bees. These records indicate that honey and wax were eaten as far back to the beginning of time. Israel was home to an apiary containing empty hives and round bees dating back to this time. This century, Greece has been home to hives constructed from earthenware. The discovery of rock art in Spain around this time shows someone moving up to a hive that is original to extract honey. This ransacking technique continues on up to the present, in different societies and with different bee types. Egyptians have been shown to have smoked hives and collected honey. Between the 1500s and 1800s, individuals used gums (tree trunks), wire bushels (skeps), and ceramics to make hives.

People encouraged bees in the crowd to make more hives to create more heavenly honey.

Europeans had to be partitioned in their beekeeping methods in the nineteenth century due to limited materials available for building hives. They used horizontal beehives in southern regions; upstanding logs were used in northern areas; straw skeps were used in the northwest because of the lower number of trees. Today's bee boxes have made it possible to give more thought to space. Two bees, walking consecutively on brushes and based on their outlines, have around 1 cm between them. This allows the bees to use that space to be free from wax.

Bee proliferation

Current-day beekeepers owe a lot to American nonconformists who helped beekeeping gain hold and grow in a nation that did likewise after American Revolution. These inventive men and their endeavors have had to suffer and they are still famous today. Langstroth developed a top-passage structure for beehives in 1851. He had a good understanding of how bees use bee spaces. He discovered that the bees would occupy the space beneath the top if it is less than 1cm. The bees will use propolis (bee stick), if there isn't enough space.

Langstroth had already protected his structure in the next year for a versatile edgehive: A crate with ten sides and perfect space between the edges. His plan was popular in the United States, Canada, as well as Europe. Langstroth's name continues to be associated with this type of hive despite several adjustments to it throughout the years.

Langstroth's honey yields were higher and Langstroth had more to offer beekeeping, making it a lucrative business venture. Beekeeping advancements exploded from the mid-1800s to mid-1900s. Moses Quinby made a great invention. Quinby invents a smoker with cries that allows beekeepers better manage their bees. The first structure was a work in progress. Producers tried to modify it, from changing the handle for less consumption to adding airspace to keep the smoker lit longer. But beekeepers have Quinby to thank. First, beehives that were human-made were simply containers that were flipped around. These hives were called skeps and had been used as the standard for over 2,000 centuries before Langstroth's creation.

Pioneers of Beekeeping

Quinby was trying to make beekeeping and honey production a profitable endeavor, but westbound extension and railroad transport gave John S. Harbison -

an early California beekeeper - the chance to send huge amounts of boxed bee honey across the United States. As trucks became more common, and beekeepers gained greater control over item transport and planning, beekeepers began to understand that they could move their honeybees north to make more honey, or south again to overwinter. It wasn't long before ranchers saw the potential of bees as crop fertilizers and paid beekeepers a fee to help them get the honey they desired.

With industrialized agriculture came a growing need for more bees. Gilbert M. Doolittle combined the best of all known strategies to create a framework which could produce mass-delivering sovereigns. He enabled huge scope sovereign creation. Although it started in the northern US, they moved south over time, as they had the advantage of a previous spring start. Soon delivery bees were the standard across the nation, and the rise of the bee bundle business was a result. Bundled Bees quickly became a household name, with Sears and Roebuck selling bee bundles as part of their shopping lists.

Amos Root founded fabricate and market beehives as well as beekeeper hardware. Some of his magazine subscribers would stay in touch with him about one.

He'd buy the rights to their creations and make a few modifications to the structure before mass-producing and selling it. Root's remarkable and amazing grandson still runs the Bee CULTURE Magazine distribution. His organization survived despite all. Root is honored by Medina High School's games groups in Medina.

Why should you become a beekeeper

There are many reasons that people take up beekeeping as a hobby. Some want to harvest honey, and others desire to preserve bees and the environment.

Sparing BEES (AND ALL BUGS) doesn't require you to be a Sparing Bee. However, all insects are at constant risk from sickness, bugs, natural and artificial elements, as well as bugs. For their resilience, honeybees are responsible for fertilizing 90% of US crops. Our reality may also be at stake if all honeybees disappear. You can help make that a reality by being more proactive about helping bees thrive.

Pollinating Your Garden

You might see natural product trees develop on your land or in a neighboring property. If you have a nursery, it is profitable for the bees. One of the benefits of beekeeping is the possibility that your vegetables will become more diverse and delicious the first year. You might also find that your bees help to develop flowering plants in infertile places. The plants may then continue to flourish as long you keep them tended to.

Eliminating Honey

A settlement is the best way for you to have pure honey. Honey is derived from the nectar taken by bees from plants. You can even plant specific flowers to make your honey more fragrant. Honey in spring might not taste as good as honey when harvested.

You can also reap and extract wax, nectar, and dust. If they are better protected from harm, bees will become more resilient to illness. And you can help others discover how beautiful beekeeping really is. Bee clubs are found in all 50 states and throughout the world. Most have helpful and informative open doors for new beekeepers. Club members often receive beekeeping assets from bee clubs, such as beekeeper books and beekeeping books, or credit from the club extractor. I strongly recommend joining a club for beekeeping and finding a guide to assist you. This will allow you to learn from someone with more experience than yours.

Urban Beekeeping

Beekeeping has been in the news a lot lately. That flood is also a reminder of the long-held enthusiasm for urban beekeeping. You will be able to make more progress if you realize that urban beekeeping has more limitations than rustic. Be a good neighbor

1) Have your neighbours look at your arrangements. They'll understand what's happening on your terrace.

2) Place a 6-foot security barrier or support around the hives in order to prevent bees flying over your neighbors' yards.

3) Keep your bees happy by pointing hive doors away from areas frequented with human traffic and setting your honeybees far away from pet areas.

Pick the right place

1) It is ideal to place hives on private property away from high-thickness open areas. Get permission if the property is not yours.

2) Due to the limited urban space available, some people may need hives in various spots, including downtown housestops. However, it is important to consider the security of wearing a suit which has reduced permeability, and whether you are occupied by bees. Even though you may be capable of carrying light or empty hardware up a stool, shouldn't you consider the possibility that you will have to transport an overwhelming beehive filled with honey

and bees? Consider every scenario possible before you try to find a hive that's in a less accessible area.

To protect your bees from their neighbors' yards, make sure you have water sources that are easily accessible and plants for scavenging.

Know your local laws

1) Consult with specialists from the city or district to determine what you may be able and not be able accomplish in an urban setting.

2) Look for a nearby beekeeper or association to learn more about the activities that are legal in your particular city.

3) Determine if there is any nuisance laws in your area which may prohibit you beekeeping at certain locations or circumstances.

Take advantage of local resources

Get involved with a local beekeeping organization. These individuals can be a tremendous asset in helping move a colony or finding an area for elective breeding. You should be prepared for possible problems

1) While swarms may be occasionally protective, they can sometimes appear alarming to neighbours. Assist in swarm prevention and remember to leave a few extras in case you are needed.

2) Openly caring for honey syrup or honey can result in battling and cautious actions. It is smarter to look after your bees while they are still in their hives.

The robber screens contain guarded conduct and have a reinforcement strategy in case you have to move the hive. This could include moving a hive off the ground or taking a hive out of the ground.

You can water urban finishing to provide your bees with nutrition in case they are starving.

Rustic Beekeeping

Rustic gardening is now well-known. This means that more people depend on bees in these settings for fertilization. Because you may have already made arrangements to give bees what is needed for their survival, country scenes make a great place for starting beekeepers.

1) You might have more opportunities to do certain activities if you live beyond city limits.

2) Get advice from other country beekeepers on the best way to run a ranch or do their work in a different way.

The right environment is essential

1) Bees need a wide variety of scavenging places, so make sure you have plenty of them.

2) A flower nursery will offer your bees different opportunities to gather nectar and dust.

3) Develop something all year to provide food for your bees throughout winter.

4) Don't disconnect your hives.

You should have the possibility to view every one of your bees from your windows. Criminals may not be able to see your hives from a window in your home. Even if your hives look similar to hives they will be less welcoming for those who are looking for them.

1) Register your honeybees with a nearby spot so you will be informed when splashing has taken place in territories that allow for extensive rural farming. It will be

easier to avoid any damage to your honeybees. You can also make sure your hives will not be affected by crop tidying.

2) Keep track and keep records of the climate around you. Also, note what predators are prevalent. This way you can ensure that you're prepared for any problem or potential trouble.

Beekeeping and a responsible spending plan

Before you commit to expensive beekeeping hardware there are economic alternatives. These economic approaches ensure that you don't contribute more than necessary until you determine if beekeeping is right to you.

A few beekeepers have the ability to manage bees. These include building hives with scrap wood and getting swarms. For successful beekeeping, parity is key.

Knowing where you can ration and what you can't is crucial.

Your gear can be built yourself to save you money. Simple tops and bases can be made for Langstroth honey hives. However, edges and containers need not be exact. Top-bars were also intended to be done by hand. These top-bar units, as well as financial plan grade Langstroth box and Langstroth boxes, are often sold for less at beekeeping flexible organizations than the wood needed to make them. Your cash can be better used for other beekeeping activities if you are willing to spend a bit of time using paste, nails and paint.

Honey stores will need extra space if they are developing. You can concentrate honey from a territory that has abundant honey and then return it to the same season. This will spare you the cost of the case and outline and save you the bees the effort of drawing new brushes. Do not neglect your hives. Fix them at the most trouble spots. If you allow hives' components to be carried by wax moths or if hardware is something you are familiar with, it will result in more cost than if you take care of your speculations.

You can build your own hives or use items from your family to make hardware and bee suits, but the most important aspect in beekeeping is purchasing bees. The best way to spend your cash is to buy better bee stock, or purchase substitute bees. The cost of beekeeping should be less than the other expenses. On the off chance of having poor bees that all die in a year, there is no need for hives or other equipment. You'll make quick profits by investing in the best bees, sovereigns, and other trusted providers. Ask for hardware needed for any undertaking in late harvest and late-fall because significant beekeeping flexible organizations usually have gear placed in a raincheck for a short time in the spring when they are at their busiest.

The Beekeeper's Toolshed

While beekeepers may have different needs, the most important tools are those that you must claim. These are the essential items that you should start with. As time goes by, you will add other less important gear. You can make your bees' lives easier by using a smoker. A smoker is a vital tool for beginning beekeepers. It produces smokes that allow quiet bees and covers to warn pheromones when a hive gets upset.

1) Heavy sheet steel for the body

2) Durable interior and exterior parts

3) Effective planning, with an adequate roar

4) Lid with pivoted pin (instead a creased steel piece) and a bolted round wire handle

5) Large, outwardly facing firebox.

Bee Brush

If your smoker leaves before you close the hive, or if you need to quickly move the bees away from the main path to close it, a honey brush is a useful tool. The bees won't take care of the bee brush so make sure it has standard fibres.

The individual experience of a smoker will quickly reveal how important it is. One time when you think you don't need a smoker because you're only going to open the beehive for a few seconds or because it's urgent is when you'll start to learn about a difficult exercise that stings in more ways than just how flighty they are.

Hive Tool

Once the bees have sealed the hive with the propolis (the bee stick that also prevents disease and contamination within the hive), the hive device will be needed to access the boxes and under edges that have been covered with the propolis.

Additional Tools

You can use the hive smoke, hive device, or bee brush to help you in all situations. But there are other tools that may be more useful.

1) Frame grasp : Used to grab an edge of a topbar hive, and pull it out from the hive.

2) Hive bearer. Placed above the highest point on a stack of hives to enable two individuals lift a full set of boxes.

3) QUEEN-cut: To catch and hold a sovereign, you can use a plastic model of a

reasonable size to more easily observe the sovereign.

4) QUEENEXCLUDER: This prevents the sovereign from producing honey by laying eggs near your hives. The J-snare, which is a hive device, can be used to lift the outlines of a Langstroth.

Hive device

Frame spacer. This makes it possible to have more space between edges.

The Beekeeper's Closet

Beekeeping can be made more appealing by choosing the right personal security apparatus. Once you have faith in your protection, you can be more focused on your bees than worrying about potential stings.

1) Shroud

2) Gloves

3) Individual Protective Gear

It is crucial to feel safe when you are in your beeyard. Standard assurance gear covers your head, your body and your hands. But you can also get a suit that will protect your legs. You can buy a jacket and shroud together or just the cover. Look for a suit that is comfortable, ventilated and durable. There are many options for suits. One-piece or two-piece suits may be easier, but two-piece suits will last longer. If you need to bend around to do specific tasks, two-piece suits will work well. A suit should be one size larger than usual to allow for more flexibility and development. Ventilated suits prevent stings while keeping you more comfortable in hotter environments. Even though it is easier to beekeeper without gloves, beginners should be more friendly with bees. Learn about the beekeepers' practices before giving up gloves.

Sometimes, being stung once or twice will help you adjust to the pain quicker.

Material, elastic, and calveskin. Check out a few gloves to determine their ability and adaptability. If you want your gloves to do more than protect your hands, they are necessary. A few people enjoy wearing several sets of Nitrile Gloves. They are more comfortable and allow for greater manual smoothness. They can heat up and make your hands sweaty. However, they are very easy to tear.

There are many options for cover types, but the most common ones are a collapsing cover or a cloak or head protector (also known SQUARE shroud), around and Alexander covers, and a fencing and hooded cover. A few covers come with a working in the cap. You can pull other covers over the top of a cap. A round cover makes it possible to see in 360deg circles. However, it might be difficult for you to see the ground ahead of you. Many full suits are accompanied by a hooded shroud. You can wear them with a baseball hat to keep your face from being covered by the face cloak while you work.

The cover is preferred over a full bee suit. You can even use it if you need to remove your cover quickly after working for a while in warm environments. Your shroud will be a good match if you give it a chance in the bright sun, in conceal, and with your glasses if you are wearing them.

The Langstroth Hive

While there are many different hive configurations, the Langstroth is the best. Langstroth outlines stop bees contacting hive dividers or casings with their brushes. They also make it easier to monitor the edges.

How to choose a box size

When choosing Langstroth bees, it is important that you choose an 8- or 10-outline width. The two sizes typically cost approximately the same amount. However, 8-outline honeybees are more popular than 10-outline hives because they aren't as heavy. You won't have problems with any other segments, regardless of the size you choose.

How to Choose a Hive depth

Each hive has four depths: deep, medium or shallow. The most recognized boxes are the profound (95/8 inches) and the medium (65/8 inches) because they allow for a lot more space for the bees. Brood rooms are most commonly made out of profound boxes. In some cases, these boxes are still called brood containers. For honey creation, medium boxes are commonly placed on top (prevalent). These are known as SUPERS. You can also use shallow (53/4 in) and brush honey

(45/8 in) boxes. The bees won't mind whatever size of container you use. They will use it for their brood, or to store honey.

The Beekeeper's Bottom Boards for Picking

Scratchpad:

Producers have different widths of 8-outline enclosures. They also differ in how they use bee space to isolate the top and base of their cases. If you are able to find boxes that belong to a single producer, it will be the best fit.

If you're arranging your honey bees, think about how big the boxes might get when full of honey. A 10-outline, deep box filled with honey measures up to 90 pounds. An 8-outline medium box is loaded with honey measures around 45 pounds. Each of the eight-outline mediums are used, which means that each part can be traded.

The base sheets have always been noisy, but screened ones have been developed as part of an IPM (incorporated nuisance the board) technique to allow for ventilation in the beehive. They are important to those who use screened bases sheets that have gadgets for catching beetles and checking them. Screened basesheets are more costly and more delicate than stronger base sheets. Large apiaries using strong base sheets recognize that the bees are able to control ventilation without the addition of an opening for a screen.

Either a temporary spread or an extending one. They are used by beekeepers of enormous scope because they allow them to place hives nearby one another on beds. This gives them the ability to move those hives quickly from one dusty area to the next. Transitory spreads don't require more than a level of wood in front, but they do have a projection so that it is easier to get the honey off. These spreads can be stuck down with propolis at the edges and to outlines which makes them difficult to expels. Specialists frequently use an extended spread that has a marginally greater amount of wood than a honeybee body. This allows for the spread to "telescope", or extend, over the crate. This type of spread requires an inward spread. At the moment, bees will pastedown the inward spreading with propolis. It is meant to be easily broken.

Utilizing Entrance Reducers

Passageway reductions are important to lower the size of the first honeybees you need to protect. Reducers made from wood can have a small opening as well as a huge opening. When the small opening becomes blocked, open another hive and turn the reducer to the larger one. A lot of beekeepers use bits and pieces of scrap wood to accomplish similar goals. The process of selecting the edges and establishments to build your Langstroth beehives can lead to a lot more dissatisfaction for new beekeepers. This easy guide will allow you to make the right choice for your hives. Select your establishment first, then choose your edge type.

Since it's more economical than wax establishment, plastic establishment is the most simple and straightforward way to get started beekeepers. Plastic establishment in wooden cases and plastic establishment in plastic edges offer parts that can withstand delivery and maltreatment. Plastic establishment can only be used by the bees as a wax setting. If it has beeswax, it will be easier and faster to be brought in brush. The plastic edges and the plastic establishment are a one piece plan that does not need to be gathered, but breaks in the casing can conceal bothers, such as tiny hive beetles.

If you are using a plastic establishment, and gathering your casings together, pick a GTB top bar (GTB), as well as a GBB base bar (GBB), for the edge type. Once you have an edge, the establishment will fly into the casing. Because plastic establishments in wooden casings can be sold fully assembled, there is no need to worry about hiding areas in one-piece plans that could harbour vermin such as the little honey beetle or wax moth. If you are purchasing unassembled outlines or establishments, ensure that the establishment and casings are arranged in the correct sizes. A deep establishment cannot fit in a moderate edge. Also, parts from different manufacturers won't fit in a medium edge. It is ideal to arrange from one producer. Casings to be used for rambles, and edges to be used for labourers (yellow).

Frames and Foundations made of Beeswax

Over 100 years ago, beekeepers had used unadulterated honeywax establishments that were placed in wooden edges. Although this is a standard decision, there are some drawbacks. The wax establishment can break in cold temperatures and stick in blistering heat. Furthermore, it can be easily damaged by misusing. Beeswax establishments often have a variety of confusing casing types. But wired wax with snares is the best choice for learners. Vertical wires are used to guide the wax. The snare, which fits under the wedge at the top bar, supports the establishment and prevents it from falling. Cross wires, as well as those that run under the top bar, will prevent establishment from falling and aid in extractors.

A mixture of wax establishments has outlined the edges with a wedgetop bar (WTBB) and a split-base bar (SBB), as well as with an edge with a WTBB and a notchbase bar (GBB). The wedge is then removed at the edge's highest point. Once the establishment has been introduced, the wedge is stapled or nail set up above the establishment to ensure it stays put. An SBB refers to establishments that are too deep or too thin in order to allow the bar to hang. To allow for cross-wire hanging, end bars must have gaps. GBB requires a perfect fit which only a few companies can ensure.

Wire Frames and Foundationless

Foundation-less casings can be used for those who require a second choice or want to avoid the inconveniences that come with wax establishments. Only a few producers offer an edge that has a V shaped wedge at the top. However, many beekeepers prefer to use the wedge as a guide by sticking a Popsicle stick (or something similar) in the top-bar. However, the wedge acts as a foundation for bees to begin assembling the edge. A foundation less case is placed between the lines of the drawn bristle. Bees usually draw it straight, but beekeepers need to ensure it is straightened by bees. You might need to mediate to make sure it stays straight. There is no establishment for labourer measured brood so the bees may have foundation less edges.

Attracting Swarms of Bees to Your Hives

While welcoming large numbers of bees to your colonies is an exciting opportunity, they are vital for progress. The best way to increase your ability to attract honeybee swarms is to plan where you want to put them.

A setup state is where the honeybees have established a home. They'll be building their nests and protecting their honey stocks. A lot of them are sometimes transported for province multiplication. The honeybees are always hungry and very meek, even though they are constantly searching for homes. Many property owners in hurry misidentify set-up settlements and call them swarms. In a matter of days, a multitude can make a passageway through space using dividers or trees. Multitudes could be as small or large as softballs. Furthermore, just like states, not every multitude will act or appear the same.

Swarm or Bait Hive Use

Regenerative amassing takes place in the late-winter and continues into late-spring. This is the best time for a large number of bees to move into an empty hive. Your goal should be to make the hive attractive for bees. The hive should be dark inside, water tight, and of the right size. An 8-outline Langstroth Box can be used. Fill it with old broodbrush and you can get a circle to place over the passage. You won't always get bees to visit your hive but the more you place them in different areas, the more likely you are to have a few.

Utilizing Swarm Lures

Swarm draws can be purchased economically, although lemongrass oil may not work as well. A few drops of lemongrass oil in a paper towel placed in a sandwich sack and left partially open inside a trap box will result in a moderately strong first oil smell that should last about half a months. Another good bait is old broodbrush. It attracts wax flies. It attracts moths by using old hive boxes or beeswax to clean the dividers.

Precautionary Steps

Multitudes of bees are a delightful way to get them, and can even help you secure hereditary qualities. It is important that you screen new hives for potential problems in Africanized honeybee-producing regions. Multitude sovereigns are often from previous seasons and bees might slaughter them to supplant her. Therefore, you should ensure that the new state constructs sovereign cells. All bees appear modify their new home. Bees could be affected by strong scents caused by new paint, synthetic chemicals, or sealants. The attempts to substitute lemon oil or essential oils for lemongrass have proved futile. While honey will attract bees to the lure hive, this is a looter taking the food to his home instead of scouts in search for another home.

Multitudes won't always get in a certain shape because they don't usually get it that way. Their structure is unquestionably noticeable. Many beekeepers believe that non-domesticated colonies are indicative of superior survivor genetic qualities that were untreated and managed for a long time. It is possible that these large numbers may be the result of vigorously treated hives.

Swarm Management

Get your bees! This is a great way to learn about beekeeping. This strategy may not be the best one for everyone, but it will make you feel comfortable catching bees. Diverse Swarm Locations. There is no way to prevent large numbers of bees from gathering. However, if you are equipped with the right hardware, the arrangement will provide comfort for all involved, including yours. These tools will work in most cases.

1) Protective rigging

2) Hive smoker

3) Eight-foot stepladder

4) Cardboard/core (nuc) box

5) Lightly colored bedsheet

6) Bee brush

7) Pruning shears

8) Queen cut

Frames

10) Drawn broodbrush

Before you go anywhere with your crate of honeybees, be sure to have a sovereign. In the event that there is a sovereign in the container the bees might remain at an open area with their tails visible. If this happens, they will fan their pheromones to attract new hives. Many situations include ground-level tree trunks; tree trunks that are accessible via a stepping stool; other human-made structures, such as a divider, post box, fence or fence; and even the ground, where dead sovereigns may be visible.

Instructions to CATCH a SWARM

1.Clean your defensive rigging. Place a bedsheet under the multitude.

2. Put your smoker on to keep the bees quiet. In the unlikely event that they are too numerous to fit in the container, you can use your stepping step stool to help them move. Give the branch a few shakes. The container should be empty by now. You can then draw in the rest of them by placing the case on a sheet near the cluster. (You can also use a beebrush to rub them onto a bit old broodbrush, and then place it in your multitude container.

3.Once you have collected as many bees and crates as you need, cover the container. Leave enough space to the side for the bees. If you're using a nuc box for transport, add everything to the case except one endbar.

Shipping and Installation of a Swarm

If the bees won't leave the container for more than 60 minutes and it is warm outside, you can create ventilation gaps by putting a few jabs in the crate. After the catchbox is shut, the bees should be as comfortable as possible in your vehicle. Next, take the bees home to establish a hive. In the event that the bees are in a cardboard/transitory bag, you will need them to be moved into a new hive within the same day or next morning. For the bees to live in a permanent home, they will need it. A freebee without an outline is more likely to heat up. You can also find the sovereign and place her in a cover or excluder for a few days. Queenless masses aren't too extraordinary, and being able to identify whether a multitude has sovereigns will dramatically increase your chances for successfully introducing them into another hive. To help the bees chill off, provide them with a 1:1 sugar to water mix and nourish them until they are ready to move into their new home.

Sovereign clasp

These bees found a owl box that was left behind and have made it their home. Keep an eye on the cage for a few minutes to check if any bees are still inside. In the event that the sovereign is not in the case, the strays may move to the cage once they have realized their sovereign has abandoned them. The bees will return to their branches on the off chance that the sovereign is still present.

Allow the bees time to go through the entire process. Then, use your beebrush for kneading them onto a little broodbrush. You can transport the bees when you have at least one crate. Brood brush will also help you to distinguish the solid bees from the wounded bees.

Different Techniques

Take a tray to help the bees get into a holder if you need to ascend a ladder to reach the multitude. Use a beebrush to delicately drive bees in the case if you have to exterminate bees that are not from trees. You have three options to introduce honeybees into your new hive. First, you can purchase a package of honeybees that includes a mated ruler; second, you can purchase a nucleus state (or core) of bees from their mated sovereign and then buy the full-size state of any bees they had previously brought into their hive. Bee bundles typically weigh 2 to 3 lbs and are often sold by higher-ranking sovereigns. The best bee bundles contain between 10,000 and 12,000 bees. They are usually working drones, automatons, and a mated sovereign.

These bees will be kept apart until they are acquainted with other bees.

It doesn't matter what hour it is when you bring bees home. The bees enjoy daylight and flying, using the sun and light to control them. They don't prefer to fly in darkness, and don't mind getting upset if things get dull. Favourable circumstances

1) Can be introduced to a hive-box

2) Keep wax brushes clean and free from any pesticides, diseases, or other bee-related substances.

3) Will be completely bloodless for the first few days after introduction. This will lessen parasite checks, and make it easier to evaluate the effectiveness of vermin treatment.

4) Have been frequently given an antitoxicant in their syrup to eliminate nosema, and loose bowels.

5) Let new beekeepers see how their bees create wax and lay eggs.

6) Can be delivered (climate, zoning allowing).

7) Gentle nature; bundled honeybees have very little to lose and no guarantees

Inconveniences

1) Unknown if sovereigns have been rated for acceptable creation/laying designs

2) Queens are not near or familiar to your locality, which could result in additional costs if you requeen later.

3) Delayed Province Development because the sovereign does not have sufficient space to lay. Bees disappear faster than they're being formed and new states dry around a month.

4) Sugar water needs to be maintained until the hive grows or there is an acceptable nectar stream outside.

Make sure you have your hive set up and in place before your wrap arrives. You must make sure your hive box is level. Also, remove all tall grass and weeds from the area around it. Just when the bees are huddled around the syrup bottle in the exact same manner as the sovereign they will be set up for introduction. You will need a large container with a 1:1 sugar-to water ratio to keep the bees happy during establishment. It's possible for the bees to become desolate if it's windy, rainy, colder then 60degF and more bubbling that 95degF (or extraordinarily lessen). If the bees are in a dire environment, you should make every effort to get them out of it.

Watch the bees gather in the gathering.

If the bees are near your home or your sovereign is deceased, it's best to contact your bee vendor right away to get a substitute sovereign. You will provide everything necessary to introduce the queen and the rest of the hive as if she were alive.

The settlement can help her stay settled. They may not be able to smell her pheromones, but they will still try to escape the container. You can set them in a lessen, cool place before you attempt to introduce the pieces. To do this, spritz the gathering area with water or sugar water. The number of dead honey bees found at the gathering base should be measured. Routinely, 1 to 2-inch deep is the standard for progressively arranged bees. In any event, 3 inches can indicate an issue with your bees.

Package Bees

Before you present, be sure to review all guidelines. It is amazing to have someone with you to guide you through the process. On the off chance you must smoke your bees, start by lighting your smoking pipe and wearing your carefully selected clothing.

1 REDUCE HAIVE ENTRANCE WITH AN ENTRANCEREDUCER: A dab with cut wood, grass or plugs. A tiny state can ensure a much better method for just 1 to 2 inches.

2 REMOVE THE QUEEN BOX AND CAREFULLY DISPENSE THE SYRUP.

3 BRUSH BEES OFF THE EXTIRE OF THE QUEEN BOB, and examine your sovereign for signs of injury and movement. You should ensure that all escortbees in the sovereign isolation territory are not dead. In the unlikely event that they have died, ensure their bodies don't prevent the sovereign from leaving.

4 REMOVE THICK CORK AT THE END OF THEQUEEN CAGE. This is in the same way as white treats which prevent the gap. Close the end of the QUEEN CAGE with a connection, but no sweets. In the unlikely event that your sovereign pen contains no treats, then you can remove the two plugs and return three more days later to expel it yourself. By hanging the pen this way, dead escortbees won't be able to destroy the sovereign exit.

When to Smoke Bee

You can surprise yourself by squeezing a bee while it is setting up. It releases a pheromone which makes other bees more empowered and more inclined to sting you. You can inhale the scent to protect yourself from other bees.

1HANG QUEEN'S CAGE--with the pastries ends up--between the two edges of the new hexagon. Ensure the bees approach to the sovereign isolate zone screen so that they can deal the sovereign and spread her pheromones through the hive.

2REMOVE OUR FOUR OUTSIDE FRAMES. Once the time has passed, bring the bees inside. You can remove the cardboard covering the opening and let the bees slither free. The bees will circle the sovereign, and begin to look around. Spread the spread out on the hive.

3COME BACK ON THE NEXT DATE TO REOPEN HIVE. To do this, take out the unfilled bag and attentively refill the four lodgings. Close the spread after a short period of time. The sovereign assertion can be checked in two to three days, or you may wait for several weeks before checking.

Just when you check the beehive again, the sovereign must be free from her sovereign breaking point. The lodgings can be inspected to determine if the bees have made any wax. This will allow you to look for eggs and hatchlings. The bees should attract the wax in 2 to 3 brushes. This is before the end of the significant week. You can keep the settlement running at 1:1 sugar-towater ratio for up to a month. Your settlement won't start making if your honeybees begin to store the sugar water inside all the open cells. You should give your bees a quart of sugar water every 3 to 4 days. They may only drink it in a few hours, or even days. A nuc (inside) bee hive has four to five edges of pulled-in brush. It contains bees in different stages, such as egg and adult. Nucs made of bees from different settlements can be sold to explicit affiliations. Talk to your seller about any concerns and questions before leaving the apiary.

1) Removed brush from established state

2) Acknowledgement and laying of the sovereign

3) All things Brood

4) Understanding the sovereign's laying scheme before leaving the Apiary

5) Most commonly, some honey and some dirt in the nucbrushes

Disservices

1) Winter accessibility in the late season means that bees could struggle to endure it.

2) The comb might contain parasites or diseases.

3) Can get out of their confinement and be inclined to crowd.

4) Unlikely to harvest honey in the primary season

After your hive is prepared, you'll be able put on your defensive dress and light your cigarettes to introduce a nuc. In the unlikely event that the sovereign has not been discharged, she will be placed inside a sovereign enclosure.

1 REDUCE HIVE ENTRANCE - Use a bit of wood or grass, stops, or an sovereign excluder to reduce the number of crawls.

2 REMOVE 4-5 FRAMES FROM the NEW HIVE BOX. Smoke the nuc first, then open the top. Next, gently smoke the uppermost point of the brush casings.

The advantage of purchasing a fully-equipped hive is that you can get started in beekeeping immediately. A built up hive normally contains 7 to 10 brood cells, usually related bees, that, as of now, work perfectly together. It will also likely to have accumulated assets in one portion of the brushes, such as honey, nectar, and propolis.

1) Can be purchased anytime because the hive's territory is changing

2) Possible option to harvest honey during the primary season

3) Less time spent worrying about a fully developed settlement

4) Understanding the sovereign's design for laying before purchasing

5) There is the possibility of wax casings becoming contaminated with pesticides, bugs or other illnesses

6) Too stressful for a heavily populated province to move to another part of the country

7) Larger settlements can have more assets to protect them, which can increase their vulnerability to being upset.

1 PLAN THE NUC FRAMES carefully with bees in space in the hive. Pay attention to not crush sovereigns or bees. If you are using a casing feeder to feed the bees, make sure they are not too close to the casing feeder.

2 CLOSE YOUR HIVE. But, if you see bees still in the hive leave it as-is. They will find their new home eventually.

In 3 IN A WEEK you can check for eggs/hatchings. If your sovereign doesn't have eggs or hatchlings within 14 days, it won't be able to survive the move. Ask a qualified beekeeper to assess your situation or ask your bee vendor for information about a substitute sovereign.

A Queen Bee is born

You can introduce a sovereign for many purposes. To ensure the hive continues to have a sovereign bee, you must quickly introduce another sovereign.

1 PICK-UP THE QUEENBOX: To check on your sovereign's health and ensure that she is well, you can take a look inside the queen box. A few sovereigns will also have specialist bees within the sovereign enclosure.

2REMOVE THE CORK at the end of the sovereign confine. White sweets also hinder the opening. The end should not contain any sweets and it should have a stopper.

3IF INSTALLING QUEEN IN A HIVAYE WITH A DRAWN BRICK OR ESTABLISHMENT. You can press the sovereign container into the brush, or lash your sovereign pen to the casing. You must ensure that the sovereign is not blocked from the outside by blocking access to the screen.

Acceptance

You can check sovereign acknowledgement while she's still within the sovereign confine. Watch the bees that cover her enclosure to look for signs of hostility. For example, they might try to sting the sovereign with their pen. To help the bees escape the sovereign confinement, you can use your gloved fingers. In the event that the bees attach to the confine like Velcro they won't be happy to live with you. You should wait until the bees show better acknowledgment and can be removed from the confine. You should not place a dead sovereign or enclosure in a bee hive. Instead, put it in the hive along with the state until the substitute sovereign is found.

The enclosure can either be placed horizontally with the candy up, or it can be placed on a flat surface. These lines will allow the sovereign to exit the pen without being hindered by any specialist bees.

Chapter 3: Maintaining your Honeybees

Once you have settled your first bee colonies, you will need to understand how bees think and what to do with them. It doesn't matter what your bees do, understanding the science behind their behaviour will help you to make the right choices. This section will explain how bees collect and use their resources. In this section, you'll also learn how to settle the executive's decisions and help them whenever necessary. This is the key to maintaining your state's stability and soundness.

Life structure of a Healthy Hive

When you understand that your bees are able to maintain their energy and flourish, both you and your bees reap the rewards. The following tips can help you to identify signs that your hive is healthy.

Beautiful Brood Pattern

Brood can be used to describe beekeepers with bees still in their egg, hatchling, and pupa stages. The hive will continue to thrive when the sovereign is stable and has created a decent brood structure. Find the signs of sound brood brush within your hives.

1) Broodbrushes that are tightly gathered, and near the focal point of container to prevent them from getting lost in a Langstroth beetle hive

2) Cells that have not been skipped--which means that the eggs are of similar age. This makes it easier and more efficient for working drones, to care for the brood.

3) Most cells are filled on one side of an edge. This is the highest quality level of sovereign efficiency.

There should be a reason why you open a hive. When you open a honeybee hive for any review, you are disturbing the sovereign and the province. This can add unnecessary weight to your bees, and could cause them long-term pain. Testing may be required to determine whether the honey is being properly stored. One box should be filled with honey for a Langstroth Hive. This is usually the medium honey super you've given to the hive. This will ensure your bees have enough honey in winter to withstand the harshest temperatures. To ensure they have enough honey to survive winter, you'll need to feed them in fall.

Sufficient Pollen Stores

Dust is the protein needed by developing bees. It's how they obtain all of their nutrients. The vast majority of bees' dust will end up in the broodboxes. Each case will usually have some dust on it in order to care for the hatchlings. If the bees are laden with dust or have more, they'll combine everything in one brush and put it on the side of a brood. If your bees have a smaller area that isn't developing quickly, check to see if they need more dust. You can collect their dust during long dust days. Once you are done, freeze it. If the bees start to produce random numbers through the brush you might have an ineffectively mated sovereign. Imperfect brood can be caused by many factors including climate, temperature, regular fluctuations, incorrectly arranged brood hunt for laying and not having enough imperial jelly or other nourishment. You should consider all possible options before you decide to supplant your sovereign. There are two options: You can buy dust substitutes online from bee-friendly stores, and then feed it to your monarchs dry or in powder patties. The dust patties are mixed with honey, sugar water, or both.

Propolis

They gather sap from trees and then take it back to their hives as propolis. This "stick" is used by the bees to cover their hives and to seal any cracks or clefts. Some honeybee species also use high-quality propolis. They appear to stick everything down in the hive making it difficult to do hive investigations. Propolis, which is both beneficial and antibacterial, is the best hive item.

Bees and Safety

The ideal reason to keep bees is to love them and support them. Bees should not be treated as pets. You need to try to see them as other people, and also find a way for you to protect them. A few beekeepers suggest that the best time for the hive to be opened is in the afternoon when the greater part of the foragers will not be present. It works great as long as you have a good climate. You might find it more difficult to assess in the early mornings if it is blistering there. Keep water near you, so that you don't have to worry about getting stung. It's smart to tell your honeybees when you're going to be in their area. Your bees will enjoy the honey and feel secure. A smoker will mask the bees' warning signals, so you won't be attacked by more cautious ones.

Beware of CRUSHING Bees

Use a hive or casing to transport, and take care. In the event that you drag a casing out of control, you might roll a bee. You should keep track of how they are moving so you can return them to the right place in the hive. To prevent bees from crashing under the casings, return them slowly to the hive. You can avoid interruptions by preventing them from happening. It's safer not to bring pets into the beeyard with you. However, if your pet is being pursued by bees and stings them, you may become stressed and occupied. You might lose an edge or have to leave a bee hive open for care.

To avoid getting stung, make sure to wear protective rigging so that you don't get stung or run into trouble. Bees will recognize if there are any sudden changes and will get upset if they see you moving too fast. Keep your phone at home, or in a quiet place.

The most important territories that bees attack are your eyes, noses, mouths, ears and face. They will also attack your hands if they believe they are in danger. If you are stung, get rid of your rings immediately to prevent further expanding. The bees are more likely to become disturbed by windy or stormy conditions. To free the top of a hive, use smoke. Then, move to an inaccessible area, get into your vehicle or inside to quiet the colony. While you won't be seriously hurt by wearing a bee suit, you might be annoyed if bees attempt to sting and pass on to you.

Brisk Tips

A good way to prevent problems is to use this approach:

1) Use security to avoid and keep a strategic distance away from stings.

2) Use a sufficiently bright smoke and add fuel to support it.

3) Limit interruptions

4) Move slowly, but purposefully, around the hive.

5) Keep your hive open as much as is necessary.

6) Know your cutoff point and collaborate with an accomplice if you don't know.

Avoid excessively burning, getting tired, or becoming dry, especially in hot weather.

If anything goes wrong, you can make a backup plan.

In case you become sensitive to venom, keep a spare drug on hand.

10) Locate your phone and know how to get it.

Examining Your Bees

Keep an eye on your honeybees for many reasons. Before you open your beehive, it is necessary to prepare the bees for this gentle interruption.

Lighting a Smoker

Smoking your beehive will allow you to be more attentive to their behavior and safety, especially since smoke can upset bee resistance components. A crucial first step to smoking your hive's hive is lighting it and maintaining it running.

You can't put enough fuel into your smoker for a typical youngster to make it past the main beehive. You can't get enough smoke in the smoker to keep the bees cool. When the fire is growing, it can release hot coals and howls. You don't have to blow ashes into the hive to inspect because they can cause harm to the bees.

1 INSERT A SMALL FIECE OF NEWSPAPER INTO AN OFFICE SMOKER. TURN ON THE SMOKER. Siphon the cries a few times to help lighten the fire. After the smoker has lit up sufficiently, add a little fuel to maintain the flames. You can continue to siphon the roars of the smoker as you go, making sure that it continues to smoke and that there is still a little flame.

2 PACK SMOKER IN BARK, MULCH OR STICKS. After the smoker has finished consuming, start pressing it with higher bits of sticks, mulch, and pieces of cedar Bark, letting out a few howls as you go. To

secure your hands, use your hive apparatus to press the fuel.

During a hive examination, you won't have the need to mist your entire bee yard with smoke. Let the predominant breeze float the smoke onto beehives. Once you have reached the hive's entrance, crush the cries. Take a couple of deep puffs and point at the entrance, approximately 2 feet away. This stream of smoke confuses the gatekeeper bees. It should send most bees back into the hive. The disarray will settle and the smoke will cover their correspondence. Honeybees have a sensitive vibration that you can contact before they know it.

After a few minutes, you can pull the cover off and blow a light breath into the top of the beehive. You can pull off the inward spread, and you can then give another puff to expel this layer. You can then air out the layers by taking a long breath along the edges. This forces the bees into the edges and out of the tops. As you move down the hive, continue to "puff," expel and puff. When you are done breaking apart layers, expel the top layer and put it in an area safe. Next, puff again. As long as your smoke is not on the passage side, give the smoker a shot upwind. After that, expel any casings from the downstream side. This encourages the smoke to continue floating gently over the bees. The goal is to only smoke a very small amount. When you pack a smoker properly, it should have enough fuel for most bee yard visits. In the event that your smoker is going out, you can grab some fuel and put it back in your pack.

The procedure can be reversed after you're done. You can smoke the bees gently to pack them again in each container.

Queenright Colony maintenance

A trustworthy and beneficial sovereign is essential for a beehive. She is essential to a hive. You should search for the sovereign every time you go into a hive.

Indications of Queenless Colony

Although she may be sovereign during assessments, this does not mean that she is absent. But you can look for signs that the queen is fulfilling her fundamental undertakings. As you become more familiar with beekeeping, you will see that a queenright colony has an agreeable buzz. It sounds like a lovely murmur. Queen less bees are more noisy, so they will flee from you when you review them.

Physical Signs a Queen is Less Hive

Your bees may be running all over the hive making it more difficult to find the sovereign. It is important to not accept that she has died or left the hive. Look for physical signs that indicate the province is less queen than before you think of requeening.

1) No new brood. You should inspect the hive to see if there are any newly laid eggs, or young hatchlings. Place the sun behind the brood and bring the casing to 8 to 12 inches of your face. Edge it to let the sunlight reflect into the cells. Look for C-molded hatchlings or minor eggs in the base. If you do not have queen-less state, there are likely to be no eggs or hatchlings.

2) Too much honey. If the hive becomes a substantial amount of honey on almost all edges, then something is probably out of order. Also, any eggs or hatchlings. Without a sovereign to lay, the bees are foragers who continue gathering nectar even when there is no one else to care for. You must in all cases try to discover the sovereign. In the event that you do find her, you will need to identify why she is not laying.

3) Laying laborer's: In the event that you have recently seen a sovereign lay eggs, but an investigation reveals many eggs being laid in cells or automaton cells, then this could indicate that your sovereign is missing.

Restoring a colony under Queenright Status

If you are confident that you can't uncover the sovereign, or any indications for new brood - and not top or creating brood - you have two choices:

1) Order another mated Queen. This requires a raiser being able to make sovereigns available for purchase.

2) Combine the provinces with a definitely QUEENRIGHT nation. It will depend on the season, climate and how accessible it is.

For the queen to emerge, you must give bees enough eggs and hatchlings. You need a stable hive, which can withstand losing future bees.

It is acceptable to store away sovereigns, but regardless of your ability to find her or whether you don't see any indications that she has been in that particular hive recently, don't act immediately. In order to find new eggs and hatchlings, you should check the hive once more three to seven days later. You should request another inspection if there are not any signs of a sovereign.

The Queen's Supplment (Requeening).

Every hive depends on a strong and productive sovereign for its endurance. You should perform routine reviews of the settlement to make sure your sovereign has enough space for laying eggs. Also, ensure your province has enough food to support the sovereign. You might need to supplant your sovereign if your brood is not satisfactory. This could include hatchlings or eggs that aren't fully formed. Although you can do this as other people consciously would expect, it's important to kill any sovereign that is not at this point ready for her motivation.

1 BUY THE NEW QUEEN PRIOR TO ACTIVITIES WITH THE OLD SOVIET. You can purchase this new sovereign by a confided-in reproducer. Then, examine her to make sure she is healthy.

2 LOCATE the OLD QUEEN. You can expel her from her hive and then slaughter her. To protect her, you can either squeeze or place her in a container of scouring liquid to make a sovereign pheromone attraction for a lure hive.

3 FEED THE QUEENLESS COLONY a 1:1 sugar-to-water blend. This will make it easier for the provinces to recognize another sovereign. Wait 24 hours before

presenting the new sovereign. It's not always pleasant to have a sovereign to be killed. An established sovereign that fails to recover doesn't get back up. Your bees can thrive if there is a stable sovereign who keeps on laying new eggs.

4 AFTER 24 HRS, expell the plug that is over the same gap than the white sugar treats and marshmallow ends of the sovereign's pen. Don't try to smack or remove the treats.

5 HANG OR PLACE new sovereign's enclosure at the top of the brood lines. You want the bees to have easy access to the screen surrounding the sovereign in order to care for her and get to know her better.

6 CLOSE UP HER HIVE. Return to her in 4 to 7 days. The candy should have been eaten by the sovereign bees and she should be walking around the casings. To make sure

she isn't starting to lay, you should check the brood week-by-week.

Some beekeepers have the natural habit of replacing their sovereigns every 1, 2 or 3-years. I like to inspect my sovereigns regularly and supplant them varying. I believe the best time for a sovereign to be supplanted is in the winter. Pre-winter sovereigns will be laying lots of eggs. You'll also have another collection of young bees that'll stay with the hive. They become more significant and grounded once the hive has not crowded. This includes the large number foragers that have been set during the immense spring nectar stream. An additional sovereign in harvest means that there is more honey in the spring.

Balling the Queen Bee

A method known as balling is used when working drones have the need to kill another creepy-crawly. They surround the bug with a tightwad or working drones, often twisting their mid-region about to likewise give sting. This heats the centre of the ball up to a high enough temperature to kill the creepy crawly. This strategy can be used to kill outside sovereigns if the conditions are right. Exercising a sovereign from a territory and giving the bees another sovereign without losing their pheromone could lead to them slaughtering the new one. That is why many new sovereigns are gradually presented with a sovereign enclosure to protect the new ruler from being balled.

Sovereign Piping

For communication with their state or with rival sovereigns, sovereigns use a sound they call funnelling to make a very clear sound. It's a monotonous, sharp buzz. You may hear it when you get to know another sovereign about a province. It could also sound when a provincial is trying to enlist a sovereign, or when the bees have her. If a province has many sovereigns, the central sovereign is likely to pipe. Other sovereigns will also pipe accordingly. The principal sovereign will uncover their area and will go to their topped cell to sting them. In this way, she will execute her adversaries for their royal position.

There are many causes and ways to control them

Amassing a state is just one way it duplicates. It's frustrating for beekeepers when they lose more than 50% of a province that is stable to avoid amassing. The administration skills you need to master are the ability to see and prevent issues that may lead to amassing. The bees aren't going to stop expanding their number of workers or gathering honey. They will only become more crowded if they have to move around. The warning signs that your bees emit are often the first to notice:

1) Refill your brood house with honey and reduce the amount of brood creation rashly

2) The bees will fill their swarm cells with regal jam and hatchlings, which are often located at the edges or base of the brush.

Add supers to allow the bees to place honey more easily. The bees have an ideal opportunity to store enough honey for the following winter after they split. A more experienced sovereign runs away with the province leaving behind a developing sovereign.

Artificial Swarms

Sometimes it is simpler to leave a hive before the bees swarm and not to constantly try to increase their space. It takes only a few actions and two or three tools to create a fake multitude. To create a fake crowd, you will need to take the old sovereign along with all the medical attendant honeys. It is important not to allow any sovereign cells into the new beehive. In all likelihood, they will bring forth and execute the old sovereign. Since it won't contain foragers and sovereign cells, this new hive is unlikely to swarm. The old hive will be left behind with all its topped brood, open brood and eggs, as well as some supers. This will protect the old honeybees from becoming too large as they have less queens than others and don't have many open brood. If they are not giving you another sovereign, it is possible to leave a few sovereign cell in the hive. If you're giving another mated sovereign you should demolish the sovereign cells. Then, check again in approximately five days to make sure that there are no more. This strategy works best when it is in the spring, right before the principle honeystream.

Given that sovereigns give birth 15 to 16 day after an egg is laid, sovereign cells will be visible for 6 to 8 days before the multitude of sovereigns takes flight. Nearly once the sovereign cells become full, the settlement can move the old sovereign into the new area. If you don't want to present another sovereign, be sure to keep some sovereign cell in the old hive.

Overcrowding Control

Insufficient ventilation and assets can cause congestion in hives. This can happen during spring brood generation, which can make a colony grow too quickly. If this happens, it can create a crowd. All of these things can be avoided by taking precautions. It will give your bees more space to spread the settlement. They will not perceive an undrawn establishment as available for use if their brood home edges are above honey or if a sovereign prohibitor has been placed beneath the establishment. The sovereign excluder that is used to prevent them from working undrawn establishments located above their heads is often something bees will resist. An edge of durable dust, or honey-topped honey that is too close the brood outline can cause a province to become honey-bound. Similar to this, the sovereign doesn't generally like to travel to the opposite end of the island.

Opening the Brood Nest's Sides

You can open the brood homes' external edges to give bees more space. This will allow them to lay more eggs and explore other areas. Depending on the season and size of the state, you might add an edge to the hive. Bees with sufficient nourishment and new establishment will be attracted and have the freedom to lay eggs. If you move brood inside the hive, it can cause distress to the bees. It is better to have two brood containers to allow bees to be more productively supervised. Many beekeepers divide single hives in equal parts or remove a small state in spring to debilitate benefactors hives. To prevent amassing, they also set up another settlement for winter misfortunes. Moving the sovereign and part of the bees to another area is the best option after the multitude of encourages has presented itself. The sovereign should be removed and the queen less settlement should accept the newly brought forth sovereign.

Making a Simulated swarm

To create a genuine re-enacted bee swarm, shake the sovereign and move a large number of bees onto the new edges (but not with sovereign cells). This allows more bees to fly back to the original hive. This young bee is not required to have a brood (or a single casing) and will draw brush quickly. If there is stable nectar, they can also make excellent brush honey.

Use a drawn comb or put it away

Honeybees need a lot of vitality building wax to create honey. If you take a few steps to ensure they draw their brush correctly, you can return it to them at the next honey stream. Drawing brush is how bees construct their wax cells using the facilities you give them in the cases. The ability to have extra boxes of the draw brush can help increase the honey reap's size. Nothing is more distressing than having your put-away brush eaten away by insects. After the honey has dried, beetles will not be attracted to it. Unfortunately, wax moths are a problem. They can make dull brush attractive to them. Treat the extricated brush using paradichlorobenzene. You can stack the dried supers up to five levels high.

All breaks should be taped shut. This is essentially creating a fumigation zone. Your stacked supers should be checked every three to six weeks if you live somewhere warm. Moths can get inside the supers once more and lay more egg when the PDB has disintegrated. You can let some air circulation in the search for a few day before returning them to their hives. Mothballs can be dangerous for people and bees if they are used. You will need to introduce a new sovereign if you are concerned about the health of the hive. You can replace harmed brush by a putaway, drawn brush.

Keep track

You can track medicines, climate conditions as well sovereign issues, bother issues, and other hive movements by keeping precise records. This helps you identify which aspects of your apiary worked well and which did not, especially in the event that you add hives.

Settlement Records

If you have many hives, you might number them and paint them an alternate shade or name them. This will help you keep track of them. If you're dressed in a bee suit you might find it easier to take a marker along to write notes on your hives. These notes can then be deciphered and substantiated later. Some data you may note include:

The way you keep a record varies from person to person. Some options include creating a hive, maintaining a spreadsheet or writing for it.

1) The date, and who performed the review

2) The assignment assigned to a specific beehive

3) Equipment components and their conditions

4) The state in which the hive is at present

5) The sovereign's hereditary traits (if ever known) and her creation date

6) The hive's manner

7) Do you see the sovereign?

8) Are there any eggs, irritations, or automatons you found?

9) The hive's population

Different APIs

It's possible to keep your honey bees in multiple bee yards. One note pad could be for your home apiary while the other one would be for your outside yards. Make sure your assignment framework has a place you can record it, in case you aren't the one doing the assessments. Do not attempt to assess more than 10 hives per day. You will need to take the time to assess each hive and not try to solve them all in one day.

You will be able to see what is happening in your immediate area at random times during the season. This will not only help you understand how your bees are doing. You can also use it to help you supplant or evacuate any plants, if necessary. Record the date and information about what's in bloom as you go around your apiary. You can also track the bees' visits to certain plants. This information will allow you to predict how bees behave and what their rummage habits are. It also helps you identify when bees are in need of sugar syrup, dust, or other care. It's fun to observe what your honeybees are eating when you harvest it. A blossom journal, which is a simple way to record your honey's scents and preferences, will help you discover the exact mix of plants used in it.

Different CONSIDERATIONS

1) Legal DOCUMENTATION. You should keep your buy or tenant agreement, your hardware manuals, as well as any enrollments that you have for your beekeeping company or apiary, along with other beekeeping records. Your hives should be stamped with a name that clearly distinguishes them as your property.

2) Expenses & duties: To assess your farm property charge, beekeepers must follow the requirements of your local burdening authority regarding beekeeping-related fees and costs. You can find information about the area's assessment code, including the number of bees required for each section of land, the additional documentation needed, and the recording

cutoff. In addition to the above, you'll need to track mileage, salaries, and costs associated selling honey or beekeeping supplies.

It is important to conduct regular checks on your hives, the bees, as well as their general condition. No matter what hive type or design you have, all of these tasks are similar. However, your bees will always direct how often and when you check your hives.

What are you looking to find?

It is a good idea to count how many bees she has around her as well as the number that are facing her. The more she has bees who take care of her needs the more successful she is. And the more grounded the pheromones that she uses to draw them in, spreading prosperity throughout the province. If you can see eggs and young hatchlings, then you most likely have a monarch. Check the casing to make sure that the sovereign's laying patterns are predictable. Then, add up how many cells are full of brood. Check for sovereign cups or sovereign cell.

Brood outlines should be secured with bees. This means most open brood has mature bees living in the area to take care of it and keep it at a comfortable temperature. It is possible to see bees at the edges of brood homes putting new wax on their brushes. New wax is much lighter in shading. It smells great and tastes like honey. It darkens each year as bees trample and use it. A good sign is the possibility that the bees overlook you during an examination. If the bees are flying around at you when you have an edge, or trying to escape from your body, it is a good sign. You can wrap them up quickly and close the hive quickly to keep them quiet.

Food Storage and Nutrition

For the most parts, bees leave dust and nectar around the brood outline edges. Bees normally store honey over the brood. This is called a honey crown. To protect the hatchlings, bees store dust along the base or sides of brood containers. What would be the number of different colors of dust you could see? A mixture usually means the bees are having a balanced diet. Imperial Flight A fully mated sovereign occasionally takes to the skies due to the weight of her mid region, which is loaded with eggs. She will, however fly if she has to, and then quickly return to her hive.

Workers drones care for the hatchlings by taking over a few uncapped cells. It is also worth researching them to try and determine their identity.

1) If you see any gatecrashers anywhere near or in your beehives, including ants, little honeybeetles, wax moths or hatchlings as well as cockroaches and wasps, arachnids and varroa vermin you need to take the necessary steps to eliminate them.

2) Signs of parasites or sickness include discolored hatchlings, darker gaps, or clingy and gooey cell or honey.

There are three things to worry about: 1) Deformed or wilted wing on the bees; 2) Spread wings on slow-moving bees; 3) Large quantities of dead bees or pupae on the hive floor.

4) Hive floor stains of wax could be an indication that burglarizing took place.

5) Observe how overwhelming each bee feels when you lift it by itself. Also, how overwhelming do the edges feel with

honey and dust? Or are they light and voidy?

6) Listen out for sounds in your hives. Compare their sounds with the hive's once it is opened. What you hear might indicate an interloper in the hive or a queenless one. Similar to above, if the sounds start at one edge, then move to another within a hive of the same type, it is possible to disengage any issues faster, even though bees are more likely to cause strong commotions when you get closer to their sovereign.

7) If your sovereign is channelling-- emanating a loud commotion that seems like a peep--this could mean potential clash. Virgin sovereigns funnel to various alarm sovereigns or working drones, if they are willing to fight to be recognized as sovereigns.

8) If you feel the smell of rot or an unpleasant odor, look for signs of sickness. Some beehives on a baseload can make a horrible odor, but hives usually smell sweet and sweet. Little hive honey beetles could have a thick sweet scent.

Examining your Hives

Before following these general guidelines, take the time to complete the "Lighting a Smoker" segment.

LANGSTROTH HIVES

1) Make sure you take safety measures to protect your bees while opening your hives using any instrument.

2) Make sure the sovereign doesn't live in a hive spreading when you expel them from a colony.

3) To avoid any possible injuries to bees who are walking through the brush,

remove the outline slowly from a honeybee hive.

4) Perform assessments at all populated areas in the hive.

5) Before removing the first outline, remove the second casing that is attached to a side-divider. You can reduce the space between the bee divider and the first edges by removing the second one first. This will avoid any possible injury to the sovereign, or to any bees.

6) Inspect each casing carefully as you expels it. This will ensure that the sovereign doesn't reside on the edge. To ensure she doesn't bounce off the casing, place the edge of each casing close to the first beehive. She will not be injured if she falls again into the box.

7) Stack your sides to make it easier to return the edges to the hive in a

comparable request. Ensure that your stacked edges face the same way.

8) Find the sovereign after you have returned the edges back to the hive. Check her out to make sure she isn't being hurt.

9) Once all the edges are back in the hive bring them to its focal point. Make sure to leave extra space at the exterior edges to stop burr brushing between the outlines. A edge spacer is also available to evenly circulate the edges once they are all returned to the crate.

10) Before you return your spread to the hive with the spread, blow two to three puffs smoke over the crate. This will encourage the bees into the casings.

Due to the abundance of goldenrod blossoms in our harvest stream, it can smell a lot like dirty rec centre socks. Not all bad stenchs are problematic.

The Tips of a Beekeeper

Because I have several hives and many outwards, my review is usually limited to 8-10 hives. I apply a bit channel tape to each hive's posterior. After every examination, I use a marker and a pencil to take a few quick notes. I can create progressively more detailed notes as I return to each individual hive to review my notes.

This will make it easy to see the eggs, and small hatchlings in bottoms. A top bar, which is horizontally oriented, means that the bees begin towards one side and move toward the opposite. Smoke the hive's entrance and spread. Then, expels any top bars remaining at the end that bees aren't yet using. You can then start to inhale smoke by inhaling. This also allows them the ability to alter. If they don't like it, they will start leaving the hive in their thousands. You can quickly close the hive and put the top and bars back on.

You can continue the assessment by taking about 8 to 10 empty bars off and putting them in a safe location. Move over the bars that aren't covered yet. Gradually unstick them. Slow down once you reach the bees. Start to see them bar-by-bar.

Contrast Langstroth Hives

The greatest difference between top bars and top bars is the inability to tip the bar sideways to view the cells. A bar with honey, wax, honey, dust and the like in it will be too thick and fragile to tip to one side. Keep the bar upright on the pivot and turn it over so the brush is facing down, hanging sideways, or hanging in a tangled mess. The wax brush is fragile and will break or snap if it gets hot, cold or substantial. There are always issues with hardware, even the most well-built. A few simple upkeep steps can make it easier to see how long your hive boxes will last and how happy your honey bees are. Water sprinkling around the beginning boxes, or gathering inside a bee hive on a board can increase wood decay. Decent paintwork is one way to stop or slow down wood decay. It consists of one layer of pre-painting and two layers that are decently applied outside. This gives you great protection from the harsh climate.

You should avoid using paint on inside surfaces and refrain from painting the edges between hardware surfaces. It will make it more difficult for the crates to be separated later if they are painted in these spots. If you suspect that the hive may have some paint on it, scrape it off and sand off the hive. Good paintwork is your first line protection. It is important to remove all paint from the exterior of the hive and seal any cracks or joints. While you can repair outside hive problems with wood clay, or with caulk to fix joints and creases outside the hive, it is not recommended that you use clay or caulk in your hive. If corner joints have begun to loosen in a container, consider development. Drill pilot openings and place waterproof paste into the holes. Then, use screws for reinforcement.

The majority of gear sold as "collected-and-painted" would benefit greatly from using more paint before use. The possibility that your hive base has termite harm, or wax moth injury--which can be detected by the football-shaped inward and outer dividers--you may be able to save your hive. This would mean that a deep box would become a medium or small box, and vice versa. The organism can cause a decay in your hive's base, which can lead to a loss of uprightness. In the long-term, you can cut 1 to 3 crawls out of damaged wood from your container. You can't save everything, so wood decay is ideal for replacing the damaged hive container.

Propolis as a Protectant

The majority of bees will use wax and propolis to cover and protect their cases. Most importantly, wax and propolis protect the container's edge rest and upper edge. Edge rest fix is not an alternative. Wood clay makes cruel synthetic concoctions. Metal wraps can set casings higher than intended and conceal little hive beetles. For hive soundness, the casing is essential. The beekeeper's dream of having a casing that attracts the most bees is possible, but you need to maintain those brushes. To strengthen corner corners, you can use an edge saver to fix broken top bars. It's often simpler to replace broken base bar. The use of an edge cleaning device allows for faster removal of wax from wood outlines. One is barely necessary to clean the wax. It would then be possible to fix cross wires, and even create a brand new one.

After five years of most amazing use, it is best to dispose of the older darker brood combs. Brush can ingest and hold any possible synthetic compounds or contaminants that bees may have introduced into the hive. They may have reached levels that are toxic to bees after five years. You can easily get rid of your brush by removing two casings per year from each crate. Scrub the entire look for beeswax to disintegrate it. Two casings will be needed for each year in a 10-outline container. You can then follow a 5-year rotation plan. They won't all want to be evacuated simultaneously, as it takes a lot bee vitality in order to make wax brushes. Bees can also reuse them for their honey. Drawing a brush is a valuable resource.

You can also use ordinary brush revolution to help offset vitality consumption and state wellbeing and prosper.

Top-bar hives are unique in that they offer favourable circumstances. They have no crates, extractors or stacks to be lifted, and can be viewed from a full view window. Top-bar honey bees are able to adapt and learn despite these benefits. It is possible to decrease the chances of your top-bar bee hive getting lost, especially if you are starting with a number or bundle. If you start with a nuc the bees already have brood and brush to handle; but if you give them a bundle they will be more motivated to stay.

1) Beeswax. Make it smell like home. Use beeswax on the edges of the top bars to rub or dissolve the wax. You can hang some broodbrush from another beekeeper, or even better, if you can find it.

2) QUEEN EXCLUDER. Use a sovereign excluder to cover the passageway and prevent the sovereign from flying out before she begins laying eggs. This will help the working drones discover their home and find a sovereign that they can satisfy.

3) Feeding. The bees won't have the need to search for nectar or go to the store to get it. They should find the food easy enough to reach that they don't need to work and can begin to build new lookovers for the sovereign to lay egg in.

I've put another bunch in a top bar hive. The sovereign excluder was installed and the bees flew out. They may go back, but eventually they realize that the sovereign is their home, and return to the hive.

Take a look at these Top-Bar Hive Ideas

Top-bar hives rarely use establishment or edges, although some have been modified.

You can be sure that the bees understand how to build their brush. However, in an original beehive, the look over can never be pulled out to examine and can instead be held up cross-supported in order to balance them in the hive cavity. As you carry out examinations, be sure to hold the brush securely as you move it. The new white bristles are the most delicate. They will be twisted easily if the brush is tipped to the side. These brushes were created to hang with gravity. One tip for beekeepers working in colder areas: When the wax brushes are extremely cold, the bar can become brittle and may cause them to knock. Temperatures above 90°F should be considered a warning sign. If you decide to open a honeybee hive, it is best to limit the amount of time spent on them.

Place bees in top-bar hives without brush means that they have the freedom to choose where to begin connecting the new brush. However, these decisions are not always useful for executive beekeepers. For the bees to build their nests, it will take regular and consistent inspections. You might find that they expand on more than one bar or form their brush connection in a curvy shape. This can be problematic for reviews or cause chaos. After the bees have produced one good straight brush, it is possible to use this search as a guide in future brush work. After that point, the bees have a straight guide on each side to give bee space between and will build another straight brush. You can continue to fill the brush with water until it is filled to about 10-12 bars. There's enough straight brush left to control and maintain the hive.

You can pull the bristles and cut the odd segment. This will allow you to place the broken section on the bar. Top-bar honey hives are more successful in subtropical or tropical climates than in areas experiencing long virus seasons. Langstroth hives are more vulnerable in weaker climates. When the honey super is placed over the top of the brood groups, the glow from bees rises up into their honey chamber. It makes it easier to feed the bees while keeping them warm. A top bar prevents all horizontal development. Bees can't move too far away from the group in case it gets very cold.

Be sure to check honey on the brood homes before harvest. You can place one on each side and sift through any empty brushes. The honey can be placed on either side of the brood zone to provide additional protection. This 4-foot topbar allows you to use an adherentboard (or a divider board) to reduce the functional space of the topbar. As the bees get older, you can move a divider board further away to give them more space. To allow the sovereign to lay their eggs, you can give them territory on either side of the brood. This allows them to achieve the moderate development they require without being forced to. It is the reverse in winter when state shrinks. If the bees are not covering the brushes, it may be necessary for you to pull some. You can store them securely and then give them back during spring development. You can also limit the amount of space that the bees require to stay warm by moving the adherentboard closer to the brood bunch. They will be able to save their vitality.

Solid Bottoms are better for managing your bees

The discussion on secure or screened bottoms continues as with any hive. Top-bars are one example. A screen is good if you're using powdered honey to combat varroa. However, the excess powder could slip through the screen. A full screen allows for too much wind flow in a top bar hive. This is dangerous for bees in winter and can lead to them making hard memories to keep their brood group adequately warm. There may be a decline of brood-raising at the bottoms, but the bees aren't likely to put brood into the base few creeps of a screen base.

Many top-bar honey bees have a removable base screen that covers the trash. However, if the trash falls through the screen onto it, the bees won't be able to reach underneath the screen. The screen can then become a rubbish heap, which can be used as a hideout for little hive moths, parasites, and hatchlings of hive insects. Top-bar honey hives come in a variety of shapes and sizes, but there must be an effective way to restrict the flow to prevent any unwanted bugs from entering. The passageway should be kept to a minimum for winter and ransacking, and to the start of a new hive. To reduce the number of open doors in the hive, you can place stopper gaps towards one side. Late spring is a good time to use all four doors. However, the main one should be left open in order to prevent heat escape. The top passage is always closed in winter to retain heat. You can unplug the gap if you notice that the top-bar is not working properly. This will allow bees to remove the debris quickly without having to move the rest of the brood. Consider it another way. It could be used for ventilation or free passage to a greater state.

Chapter 4: Bee Harvests

Collecting time is probably the best part of the relationship between beekeepers and bees. The beekeeper can appreciate almost everything that a honeybee produces: beeswax honey, dust and much more. It is hard to find a better feeling than the first honey that your bees harvest. This part will show you how to make fun and useful items out of hives, such as candles, skin care basics, and other items. There are many ways you can use the items in your hives. These were all created by bees who have made the most of their business.

Results from the Hive

Many believe honey is the main outcome of a hive. But a hive is capable of producing many fascinating and versatile products. Honeybees make items from the honey they gather. It is an incredible association, provided you keep enough for the state's needs while you gather these items.

Honey

The reason we bee keepers is to collect honey, is what we consider beekeeping. Honey tastes delicious, and beekeepers appreciate honey for all its benefits. The USDA gives three honey evaluations (A-B-C) in the United States. These ratings are used to assess the different rating factors.

1) MOISTURE Content: Level of water

2) Absence of deformities - The absence of particles and propolis

3) Flavours and fragrances: Taste and smell from the primary flower source

4) Clarity: Clearness, simplicity, and no air bubbles

Honey has also been assigned shading that doesn't affect the review. However, it does influence the honey's flavour. Light honey is milder and darker honey more grounded.

1) Water White

2) Extra white

3) White

4) Extra light golden

5) Light gold

6) Amber

7) Dark, golden

Honey usually contains peroxide activity. This is how honey can resist bacteria and won't spoil if it's kept dry and stored in a dry place. Honey is also hydrophobic, meaning it can retain moisture from the air around it. This makes it even more crucial to avoid getting dampness.

Beeswax

While bees use wax to store honey and protect pupae from developing hatchlings and larvae, bees can still make unique products with beeswax, such as candles and skin care items. Regal jam is delivered in beekeepers' hypopharyngeal gland to take care of two young hatchlings and the adult sovereign. The bees do not store illustrious jelly; it is constantly replenished. When sovereign raising occurs, nurture bees are able to give excess of the jam to sovereign hatchlings. Any that is not used will eventually accumulate at the bottom of the cell. Although the sovereign lives longer than work drones in terms of ovaries, there was a lot speculation that people could be made to eat imperial jam, which would increase their ripeness, longevity, and youthfulness.

Propolis

Propolis is a honeybee product. Honeybees collect sap from trees and plants and combine it with spit, beeswax, and other natural pitches. It can be used to seal holes or splits within the hive. Propolis can either be removed from the hive's edges or hive-dividers or collected using a propolis screen trap. Numerous stories have been written about the various medical benefits of propolis. One company even makes toothpaste with propolis that helps to secure and feed the gums. The dust is full of nutrients, proteins, and minerals. Many people have confirmed that the ingestion of small amounts daily of dust can lead to an increase in their vulnerability to occasional sensitivities. Some dust sustenance claims that you have seen an increase of red and yellow platelets, decreased cholesterol, and lowered triglycerides.

Bee Bread

Bee bread can be eaten by everyone as it is high in protein. It is essential for winter survival and settlement wellbeing. They collect dust from the hive and return it to it by searching bees. They dump the dust straight into the cells, close to the honey stores, and empty it in the open cells. For many new beekeepers, cutting brush is their new side hobby. Cutting brush isn't quite as time-consuming or chaotic as the process of separating honey and wax. It also has numerous health benefits. You can sell the items that you collect from your bees by using a cutbrush. The best time to freeze Dust brush honey is 48 hours. This will kill the eggs and keep them from incubating in the bundling.

This also kills the wax moth eggs that are in the drawn toothbrush. But it doesn't stop them from getting to your brush and laying new eggs. The cut brush can be stored in a more relaxing space if that is possible. It's possible to save your effort by wrapping the brush in plastic and placing it outside. If you have a province where you keep bees and are capable of parting them into another hive you can either sell it or keep it for future honey production. For a beekeeper who has a growing apiary, it's difficult to bring in the cash you need by selling your bees.

Setting up Your Honey Extraction Facility

Honey extraction can be enjoyed by all members of the family. Just a few items and you can make raw honey bottles for yourself or to give to your loved ones. You can extract honey from your own space. This will collect all the wax tops left by the honeycombs. A honey extraction process can be complicated. To prevent this, make sure your tote is at least 6 inches deep, 3 to 4-foot long, 18- to 20-inch wide, and includes a top. Warm, encased honey can stream more easily. There is also plenty of space for gear.

Extractor

There are many types and models of extractors. Smaller extractors often have hand wrenches. Larger ones, which hold at least six edged, are usually electric. Some extractors are equipped with plastic tanks. Larger tanks are made from treated steel to avoid rust. They use radiating power in order to move the honey from the uncapped cell onto the extractor's dividers. The honey is then able to sink to the base. A can with a lid that is made from nourishment grade plastic. It must be made from nourishment grade material. You can find them in 1-gallon or 2-gallon sizes at a local home improvement store. Or you can get a used one from your local bread bakery.

Hot Knife

Either you can buy an electric uncapping swivel blade or you can use one that you currently have, such a serrated bread knife. To use a bread blade, heat some hot water. After the water has boiled, dip the blade into the water to remove the wax from the honeycomb cells. The capping can be cut with a virus blade in most cases, but it's more complicated. Cut the capping in the middle of the uncapping tray and save the capping for beeswax. These devices will be used for uncapping wax cappings and discharging honey. An uncapping rolling device has spikes on the surface. However the uncapping scraper has metal tines to penetrate. It all depends on how little damage you wish to do to the brush. The screen will remove a large amount of the wax particles as well as any bee parts. Your honey can then be packaged and eaten. Some beekeepers use larger screens for their first run. Then, after that, they use a miniaturized scale channel to check the scale and prevent honey crystallization. While small amounts of wax and dust can speed crystallization, nearly all honey will take form.

1) Hot knife

2) Uncapping Roller

3) A distracting, hand-wrenching extractor

Step by Step Instructions to Remove Bees From a Super

A smoke board is a device that uses a synthetic smell that the bees don't love. Find one that will not only annoy but also not in any way hurt them. For example, Fischer's Bee-Quick. They are then pursued from the honey supers by the smoke board. A smoke board is a device that uses a synthetic smell that the bees aren't used to. It follows them from their honey supers with a scent that they don't mind.

1MAKE A FRAME is about the same width and length as a honeysuper, but it's only 3 to 5-inches down. You can cover one side using felt or fabric.

2SPRAY CHEMICAL ON THE MATERIAL, and then set it over the honeysupers instead of their top. A large number of the bees will soon leave the honey supers within 5-15 minutes.

3PULL ALL THE CAPPED HONEY SUPERS OFF THE HIVE, and then put them in plastic totes or complete the beeyard.

After you splash the honey, make sure to place the smoke board onto it. If you don't brush the bees with tenderness, it is more likely that they won't be as kind to you. You can use your beebrush, or a lot more long grass, to brush bees off each super and every casing. After you remove all bees from a particular casing, you can move the outline to a bee yard or to an encased container to keep them away from any future bee events.

Utilizing Bee Escapes

Bee escapes can be used between the honey supers and brood homes. It provides a way for bees to escape, but not back into the super. They can be used in two ways. One, place a few honey supers at the top of the bee departure. Two, stack them on top of each other and then wait for the bees to leave the brood room. It may take between 15 minutes and 3 days for the bees leave. However, you will have a few tenacious bees left to overcome before gathering.

Filtering and extricating honey

Once the bees have been removed from their honey supers and taken their honey, you need to get the honey out from the capped casings. While this may seem like a complicated process, once you get the chance to taste the raw honey, it will make you realize that all of it was worth it. There are two types: outspread and distracting edge extractors. The vast majority of smaller, hand-turned extractors do not have a common purpose. They can remove only the honey edges. To separate the opposing side, you'll need to flip the casing. You can choose from plastic or treated-steel extractors that won't rust. To keep the head dashing, you'll also need to get nourishment grade bearing oils. A honey entry valve should be installed at the bottom of every extractor.

Removal of HONEY

1- REMOVE WAX CAPTINGS. You can use your serrated blade or a knife to cut the wax cap's off and put them into the cans. This will give you access to the cells.

2- LOAD ALL UNCAPPED FRAMES INTO THE EXTRACTOR. If you have a space with a case, be sure to adjust the edges of the extractor. You should always keep the honey entryway clear when extracting honey. The casing spinner won't work if there is too much honey in the tank.

3- Use a nourishment product that has a screen or sifter to cover it. The screen will be blocked by the wax during extraction. To extract the wax from the screen, use a spoon. You can also add the wax to the wax caps.

Utilizing Extracted Honey

RUN THE CRANK ON YOUR EXTRACTOR.

Begin slowly, and then move on to compel. However you must extract as much honey as possible from the cells. The honey will fall to the bottom of the extractor. It will then slide into the container through the honey entryway. After letting the honey settle, you can return to the fixed basin. Any wax bits will rise to top. A sheet of saran wrapped wrap can be placed delicately on top of the honey can. Take care to lift the plastic away from the honey. The honey will be pulled out of the plastic and the plastic will separate the honey.

Beekeeping has a unique advantage: you can collect honey to eat, which is something that is easy to do without requiring a lot of effort. You can also share your honey to the world by using these additional methods. Food Safety. Honey is the ultimate nourishment. It never ruins. You are correct, but there are some critical components to this:

1) Honey should be kept in a locked container. This will prevent honey from getting damp. The honey will lose its strength if it gets wet.

2) Moistness can sneak into honey holder when you open it. In case honey is not used often, you can make use of smaller

compartments. This will prevent honey from sticking to your skin.

3) Honey has a pH of between 3 to 4.5. This is very acidic. Microorganisms as well as other life forms cannot survive in such an environment. This is terrible for animals, but it's amazing for us.

4) Bee catalysts separate honey into side-effects gluconiccorrosive (glucosonic) and hydrogen peroxide (hydrogen peroxide). We as a society recognize the damage hydrogen peroxide can do to germs. This is why honey offers so many health benefits.

Some honey affiliates overfilter honey that they bottle. They even remove the distinctive dust from the honey, effectively deleting the mark. It's impossible to determine where the honey is from once the particles have been removed. Because a significant amount of those particulates are dust, which has its benefits, you should limit filtration. Glass, plastic, and canning vessels are all good options for packaging honey. It doesn't matter how you store your honey. Make sure that you have a good cover. The honey will be unsafe to eat if it has been exposed to outside contaminants. You have two options: a pourramble or a pipe for getting honey into a container. Alternatively, you can use a tank or container that has a valve to allow honey to enter.

It is important to leave an air gap between the container's highest point (or the top) and the top. It's not necessary for honey to rest against the top when you aren't using it. I love the square glass Moth bottles. They use plugs to fix. To maintain a seal, be sure to clean the stopper and the jug lid. For honey sales, you need to know some details about your honey.

1) It must be named the primary item, which is honey.

2) If the central fixing of the dish is honey, you don't need a list. But, on the off chance you used other fixings in your recipe, you should include them in a standard articulation.

3) Include your contact data as well as the name, full address, and phone number of the seller (where the honey is packed) on your first name.

4) Print the net honey-load (short the container) in pounds/ounces.

Apart from these basic marking requirements, you should consult your state of residence to learn about the laws that apply to honey sales.

Honey extraction in a Top-Bar Hive

The best thing about keeping bees in a top bar hive is that you won't need to buy an extractor. This tool is designed for extracting honey from surrounded hives.

To extract and concentrate honey from the top-bar, and afterwards, to enjoy that delicious nectar.

You can also take the honey out of the beeyard and put it into a container. Then, let the bees clean the container. It is possible to create a looting scenario in your bee apiary. Be patient. The bees create brush when they need it, typically by placing brood brushes down towards one side of the hive. Later, they make honey stockpiling toothbrushes. Keep honeycombs down towards one end of your top-bar beehive to help reduce their number. I like one casing per brood. This will allow me to overwinter in the state.

Evacuating the Comb

Don't forget your gloves and shroud before you begin. You can blow smoke into the hive as you open it.

1 STARTING AT UNINHABITED END. Take out 5-10 top bar to make more space in the hive.

2 COUNT THE BROOD BARS. If you have honey bars larger than brood bar, create a new honeybar, then dismiss the bees. Rehash the process for each additional honey bar. Either clean out the cut toothbrush or place it in a container. Then, fill the container with healthy honey. Natural product presses can be used to squeeze the honey out of the cut brush. You can learn how to store honey by reading "Packaging and Labelling Honey."

3. Take each bar one at a time. Once you have finished cutting, load up your blade and begin to cut the new white bristles. Use the top bar for a guide to the bees when they redesign.

4 WITH THE REmaining COMBS, use gloves or clean hands for squashing the

honeycombs. After you have squashed the largest part of the toothbrush and honey is visible, you can pour the wax into the sifter. You can let the basin sit to deplete as much honey as you like.

Gathering wax

Honey is the mainstay of most beekeepers. But there are other uses for beeswax. To make your crude wax more usable, refine it. It takes approximately 7 pounds worth of honey to make one pound wax. Beeswax can come in many colors depending on whether the brushes are old and what their purpose was in the hive. The honeycomb wax is typically the lightest, and it is considered to be more attractive. Beeswax candles are cleaner and lasts longer than other candles because it smells just like honey. To protect your skin, you can use honey extract wax as a cap and save the dark wax from old brood candles. To use the wax, you must first collect it from your brushes. Then you need to dissolve and clean it. This guide explains how to make a stew pot.

1 Fill a TEFLON COATED ROCKPOT with water. This will raise the temperature to approximately 200oF.

2 STIR IN THE WAX CRACKED FROM THE BRUSHES. Let it heat up. The wax will rapidly dissolve, and then you can start to skim off the water.

3 Heat in a slow cooker, and let cool. After the wax cools to a thick consistency, it is possible to remove the water from the wax. You can see that energetic particles have settled at the base.

4 USE A KNIFE FOR REMOVING THE SOLIDS. This will leave you with a perfect block of wax.

You can also soften the wax using water. To do this, you need to pour the liquefied mixture through a towel and into a bowl of water. The paper towel serves as a conduit for the stable particles. If you have enough cleaned wax, it is possible to remelt it and make flame melds or small wax shapes. You can then sell it or use it to make moisturizing bars or lip ointments.

Chapter 5: Beekeeping Supplies

Dear Reader. Grab Your Bonus From The Table Of Contents Before it Goes Away

It is essential that you have all the supplies you need to be successful as a beekeeper. There are several products that you must purchase and some money you will need to invest. There are companies that sell complete kits, so you can save money on some products. Kits can be quite costly so it really depends on what you're willing to spend. Let's now take a look at how to get started.

There are four essential things you need to get started with honey production. You will need a stacked honey box, protective gear and a smoker to start your colony. You will need the hive, or stacked box, to get your colony started. The box must be made out of untreated cedar or pine wood. You can make parts of the beehive by yourself. However, if you are a beginner these parts are difficult to make.

The bee box will be made by you purchasing two different sizes bee frame sizes. The dimensions of your honey box will depend on the size and shape of the frames. The frames are the place where the eggs are laid, queen lives, and honey is stored. Buy an inner and outer cover for your frames. These items are more complicated to make, so it is simpler to just buy them. A store can also provide an exclusion. Below you will find a list of what you require. The excluder refers to the queen. You don't want the queen in the upper frames of your honey, as she is the one that lays all the eggs.

Also, be sure to buy an entrance reduction to stop pests entering the home. They will go from the top to the bottom. Important is also a rack to hold your bee box. The bee boxes you see in fields will show you how to put them together. Your beebox will have the following layers: the base at the bottom, an entrance reducer with a screen bottom, two deep beeboxes with frames, the queen decreaser, and two smaller honey bee containers with frames. I'll be going into greater detail in the next chapter about the bee box.

This list includes the items you must buy and what you can do.

Have to buy

-Inner and outer covers

You can get a range of shallow frames from -16 to 20 for your honey

Deep frames from 16 to 20 inches for honey colonies

-1 queen excluder

-1 Entrance Reducer

-1 screened bottomboard

You can do it!

-2 shallow boxes for bees with untreated pine and cedar wood

2 deep beeboxes made from pine or cedar raw wood.

-base

A colony will be ordered in advance and accompanied by a queen.

Chapter 6: Preparation and storage of the Bee Boxes

You will need to ensure that your town permits you to have the beeboxes and place them according to their rules. You will set up your boxes according to the rules for easy harvesting. Your location may need to be adjusted if you have animals that are able to get into your honey boxes. Sometimes you need to store your boxes higher.

You can also see how to set your bee boxes up from the top. This is how a beebox is set up in a field. Set up your bee box exactly as before. Starting at the top you will see the outer and inner covers. There will be two boxes or supers at the bottom. These are the boxes that you will use for honey, as they are known. There are eight to ten honeycomb frames in each of the shallow boxes. These frames hold the honeycomb. The queen excluder is next to ensure she does not get into honey. Below you will find the queen excluder and two deep hive box covers.

Place the entrance reducer on top of the screened base board. Now you are going to make your deep box. The boxes are simply deep boxes. To ensure that your deep frames can hang on to them, you will need to create a platform for the frames to rest on. The size of the frame you choose will affect its size. The queen excluder will be placed flat over the top box. Next, attach your honey frames to the two shallow supers. Next, the inner cover will be applied and then the outer.

There are many variants. I have just described one. Once you decide to purchase a starter set or make your own, you will be able to see the different options. These options were developed by beekeepers with years of experience. You have the option to make it easy or choose more complex systems depending on what your budget allows.

Once your bee boxes are ready, you can now choose your bees to go wild in their new home. They will be able to start making delicious honey!

Chapter 7: Safety with bees

Do you get stung? Yes. This doesn't mean that you need to be afraid, unless of course you are allergic. If you have an allergy to beekeeping, it is unlikely that you were meant to do so. If you were young, you might have been stung. That is how it feels: some pain. You build up antibodies to the beesting sting, the more you get stung. This is the silver-lining: getting stung more will increase your resistance. You don't have to be fully exposed. You should take precautions to avoid getting too stung. Here are some key points to help.

-Bee suit

-long hive tool that you may need

-smoker

Protective headwear for -bees

-Bee long gloves

You could use the only bee head protection, the veil and gloves. An unrefined pair of jeans in which all skin points are covered, such as jeans tucked into socks. Put a shirt on, with your sleeves tucked in. Next, put on a protective plastic layer that is durable and tucks in every part of your skin. Use long, tough rubber gloves to seal the opening between your shirt and gloves. Preparing yourself in advance is the best thing you can do to keep an eye on the bees.

The smoke is there to make you feel afraid. The fear can disrupt the bees. Preparing them for your entry into their home is as easy as smoking to ensure they don't smell it. This makes it much easier to get in and do the work.

Chapter 8: The Beginning of the Bees

Although there are many varieties of bees available, the most common ones are the Russians and Italians. You have the option of choosing which type you would like to raise. You can have each of their unique characteristics. There are two boxes you will need to add. It's your choice, and it is enjoyable to decide which ones.

Russians seem to have a high popularity because they produce well and are resistant against parasites. The Russians also keep a smaller winter population, which is a benefit for you. It is important to understand that winter honey needs to be supplied at 30 to 90 pounds. to keep just one colony alive.

For your best delivery, make sure you order your bees at least two months in advance. Bees can be very expensive. Your order will include a queen, three pounds of worker and a few other bees. If possible, try to find a local supplier. Once you have the bees in your box, you will find that the queen is in a different box. The queen should be placed aside. Next, you need to open the top and flip the box over onto the tops of the frames in the bottom-deep bee box. The bees will come out if you shake it a bit. Next, remove the frames from the top bee box and place the next one over it until the bees have settled down.

Before placing the empty bee box on top of the bottom one, place the queen in the middle of the bottom box with candy side down. In order to make it easier for the queen to emerge from her box, you'll need one frame removed. You will see a little bee candy in the bottom of the box. It is likely she will wander about the bottom two beeboxes. The piece in between the smaller ones keeps her away from the larger chambers. This allows you to get the pure honey. Once she is out, take the little box off and return the frame. After the queen is out of the little box, place the frames back into the second deep honey bee box. Now it's time to wait. Let the bees gather the nectar from flowers and create the honeycomb. Once the honey is cured, it will need to cure.

This is how bees function. One queen lays all the bee eggs. Her eggs are fertilized by males who immure the queen and do nothing else in the colony. The queen will lay approximately a thousand eggs every day. Don't worry if it seems impossible to fill all those frames. The queen will lay over a thousand eggs each day. Honey harvesting takes place at the end the summer but you don't want to wait too long. It must be still warm.

Chapter 9: How to Maintain the Hives

It is vital to maintain your bee colonies and ensure they are functioning well. Your bees will continue producing honey and growing in number. There are several things that you should watch out for. There are many things you need to keep an eye on, including the queen's effectiveness and parasites. You will visit your bees only a few time to make sure they are in good health.

Your time spent in your bee boxes must be minimal. It is easy to get curious. You should make sure your bees are happy. If you're confident that your bees are healthy and working well, leaving them alone is the best option. If you have to continue pulling out the frames, no more than four or five times is the maximum amount you can go in. While you do not want to disturb the honeybees, you can check on their hive to ensure everything is OK.

What are your priorities when inspecting the honeycomb and frames? First, get your long hive instrument. It is similar to a Crow Bar. This tool is used to separate the honeycomb from the frames and lift them out of the box. Once the frames are lifted out, you can inspect the eggs and larvae for signs that the queen is still alive. It is possible she may have been buried. If you don't see her, that's okay. You will recognize certain colors. You'll be able recognize pollen, nectar, and other colors. Green honey is a dangerous ingredient that has high moisture and is unfit for human consumption.

If the bees appear to be staring at you and line up as you remove the frame, it's a sign that they are angry. They are about swarm. The bees need to be free to do their work. Do not enter their hive.

A healthy sign that your queen is happy and working well is a small colony. Spotty colonies with empty cells are not good signs and may indicate that you need to replace the queen. Simply order another queen and expel the old queen. Workers bees will not tolerate a non-productive queen.

Chapter 10: The Harvesting of Honey

Although the exact time a honey harvest is ready varies depending on where you live, an average beginning beekeeper will look at an August harvest. While you may not get much in your first bee season, you will be rewarded with a substantial harvest the second year. However, it is definitely worth the wait. If everything goes well, your first season will yield honey. This is how you get the honey you have worked so hard to create.

To make sure your honey is ready to harvest, you need to check it. You can check your honeycomb by placing it parallel to the ground. When there is no drip, it has cured. For a successful crop of honey, you will need eighty percent or greater of capped honey to be able to harvest the honey from your frames. You must smoke your bee box immediately after it opens. It is essential to remove all bees from your frames before extracting the honey.

The bee blower can be used to remove them from the frames or a silky beebrush. After you have removed all bees from your frame, you need to have an empty cup to put the clean frame in. You will not extract the honey from the bees. You will bring the honey inside to your honey extracting centre. You should not wait too long to harvest the honey as the bees could eat it.

Once inside you will find a honey extraction tool which you will need, as well as a scratcher. You will first need to use the scratcher in order to remove the honeycomb from its sealed cap. Your frame is then placed in either an electric-powered or hand-powered extractor. The extractor spins it pushing the honey to its walls. Finally, the honey drops to the bottom. Keep it glass. It's the easiest to clean and doesn't contain any BHT.

While you can extract the honey with a standard extractor, it's a lot harder. The benefit of this method is that the wax cells can be left behind, so your bees don't have as much work. You may find them to be very grateful for this. You can use a fork to scrape the frames into cheese cloth. The honey will then drip into your container. The honey will flow faster because it is warm. It is not as strong as honey from the supermarket that has been sitting on the shelves. It is also possible to boil a honeycomb and let the wax rise to the top before using the wax for candles or other purposes.